BN 0-8373-6643-7

DANTES-43

Rudman's Question and Answers on the

DANTES SUBJECT STANDARDIZED TESTS

AF770802

Subject Examination In . . .

TECHNICAL WRITING

Questions and Answers

NLC

NATIONAL LEARNING CORPORATION
212 MICHAEL DRIVE, SYOSSET, NEW YORK 11791 (516) 921-8888

National Learning Corporation

212 Michael Drive, Syosset, New York 11791

(516) 921-8888

PRINTED IN THE UNITED STATES OF AMERICA

PASSBOOK®

NOTICE

This book is *SOLELY* intended for, is sold *ONLY* to, and its use is *RESTRICTED* to *individual*, bona fide applicants or candidates who qualify by virtue of having seriously filed applications for appropriate license, certificate, professional and/or promotional advancement, higher school matriculation, scholarship, or other legitimate requirements of educational and/or governmental authorities.

This book is *NOT* intended for use, class instruction, tutoring, training, duplication, copying, reprinting, excerption, or adaptation, etc., by:

(1) Other Publishers
(2) Proprietors and/or Instructors of "Coaching" and/or Preparatory Courses
(3) Personnel and/or Training Divisions of commercial, industrial, and governmental organizations
(4) Schools, colleges, or universities and/or their departments and staffs, including teachers and other personnel
(5) Testing Agencies or Bureaus
(6) Study groups which seek by the purchase of a single volume to copy and/or duplicate and/or adapt this material for use by the group as a whole without having purchased individual volumes for each of the members of the group
(7) Et al.

Such persons would be in violation of appropriate Federal and State statutes.

PROVISION OF LICENSING AGREEMENTS. — Recognized educational commercial, industrial, and governmental institutions and organizations, and others legitimately engaged in educational pursuits, including training, testing, and measurement activities, may address a request for a licensing agreement to the copyright owners, who will determine whether, and under what conditions, including fees and charges, the materials in this book may be used by them. In other words, a licensing facility *exists* for the legitimate use of the material in this book on other than an individual basis. However, it is asseverated and affirmed here that the materials in this book *CANNOT* be used without the receipt of the express permission of such a licensing agreement from the Publishers.

NATIONAL LEARNING CORPORATION
212 Michael Drive
Syosset, New York 11791

Inquiries re licensing agreements should be addressed to:
The President
National Learning Corporation
212 Michael Drive
Syosset, New York 11791

PASSBOOK SERIES®

THE *PASSBOOK SERIES®* has been created to prepare applicants and candidates for the ultimate academic battlefield—the examination room.

At some time in our lives, each and every one of us may be required to take an examination—for validation, matriculation, admission, qualification, registration, certification, or licensure.

Based on the assumption that every applicant or candidate has met the basic formal educational standards, has taken the required number of courses, and read the necessary texts, the *PASSBOOK SERIES®* furnishes the one special preparation which may assure passing with confidence, instead of failing with insecurity. Examination questions—together with answers—are furnished as the basic vehicle for study so that the mysteries of the examination and its compounding difficulties may be eliminated or diminished by a sure method.

This book is meant to help you pass your examination provided that you qualify and are serious in your objective.

The entire field is reviewed through the huge store of content information which is succinctly presented through a provocative and challenging approach—the question-and-answer method.

A climate of success is established by furnishing the correct answers at the end of each test.

You soon learn to recognize types of questions, forms of questions, and patterns of questioning. You may even begin to anticipate expected outcomes.

You perceive that many questions are repeated or adapted so that you gain acute insights, which may enable you to score many sure points.

You learn how to confront new questions, or types of questions, and to attack them confidently and work out the correct answers.

You note objectives and emphases, and recognize pitfalls and dangers, so that you may make positive educational adjustments.

Moreover, you are kept fully informed in relation to new concepts, methods, practices, and directions in the field.

You discover that you are actually taking the examination all the time: you are preparing for the examination by "taking" an examination, not by reading extraneous and/or supererogatory textbooks.

In short, this PASSBOOK®, used directedly, should be an important factor in helping you to pass your test.

DANTES Subject Standardized Tests

INTRODUCTION

The DANTES (Defense Activity for Non-Traditional Education Support) subject standardized tests are comprehensive college and graduate level examinations given by the Armed Forces, colleges and graduate schools as end-of-subject course evaluation final examinations or to obtain college equivalency credits in the various subject areas tested.

Active military personnel may take CLEP (College-Level Examination Program) examinations without charge through the DANTES program. Information about DANTES and CLEP may be obtained from the Test Control Officers at military installations, and from Educational Testing Service in Princeton, New Jersey.

The DANTES Examination Program enables students to obtain college credit for what they have learned on the job, through self-study, personal interest, correspondence courses or by any other means. It is used by colleges and universities to award college credit to students who demonstrate that they know as much as students completing an equivalent college course. It is a cost-efficient, time-saving way for students to use their knowledge to accomplish their educational goals.

Most schools accept the American Council on Education (ACE) recommendations for the minimum score required and the amount of credit awarded, but not all schools do. Be sure to check the policy regarding the score level required for credit and the number of credits to be awarded.

Not all tests are accepted by all institutions. Even when a test is accepted by an institution, it may not be acceptable for every program at that institution. Before considering testing, ascertain the acceptability of a specific test for a particular course.

Colleges and universities that administer DANTES tests may administer them to any applicant -- or they may administer the tests only to students registered at their institution. Decisions about who will be allowed to test are made by the school. Students should contact the test center to determine current policies and schedules for DANTES testing.

Colleges and universities authorized to administer DANTES tests usually do so throughout the calendar year. Each school sets its own fee for test administration and establishes its own testing schedule. Contact the representative at the administering school directly to make arrangements for testing.

Your Experience Can Earn You Credit.

It's a fact. People learn in many ways . . . through hobbies . . . interests . . . reading . . . work. The DANTES Program is a way for you to "cash in" on what you've learned from *your* experiences.

Take advantage of DANTES tests to earn college credit for what you know. More than 30 subjects are offered. There are bound to be areas where your knowledge can pay off academically.

Here are just a few of the program's benefits for you:

- ***Time and Money Saver***
  You don't have to sit through classes and pay for courses when you already know most of what's being taught and can learn the rest yourself.
- ***Credits Widely Accepted***
  Hundreds of colleges nationwide now give credit for DANTES tests. And the list is growing!
- ***Quicker Route to the Degree***
  Your dreams are a little closer than you thought possible. With DANTES.
- ***Start Fast***
  No need to wait to take advanced courses. Passing DANTES tests gives you a real head start.

Sound Good? Here's How to Get Going!

Check with your college to see if it administers the DANTES program. If the college doesn't but would like to, ask that someone **call collect** (609-951-1756) or write to the DANTES Program Office (see address below).

If you find that your college *accepts* DANTES credits, but does not *administer* the tests, here's what you do —

- Write to the DANTES Program Office for a list of colleges that administer the test. Chances are there's one nearby.
- Call any of the ETS Regional Offices (Tucker, GA; Oakland, CA; Evanston, IL; Phoenix, AZ; and Princeton, NJ. DANTES tests are given at all of those locations.
- In the military or in the reserves? DoD employees and family members may be able to take DANTES tests at a Military Education Office or Navy Campus Office.

Check Out the DANTES Test List.

→

Match the titles with the things you have learned. Consider how much college credit you might gain by taking some DANTES tests.

Your college should have fact sheets with more information about each DANTES test. If not, please write to us directly, or e-mail us at dantes@chauncey.com. Tell us which subjects interest you.

DANTES Program Office
Educational Testing Service
Princeton, NJ 08541-0001

from the official announcement for instructional purposes

MATHEMATICS

Introductory College Algebra
- Functions and graphs
- Fundamental algebraic operations
- Linear equations, inequalities, word problems
- Radicals and exponents

Principles of Statistics
- Descriptive statistics
- Correlation, regression, probability
- Chance models, sampling, test of significance

SOCIAL SCIENCE

Art of the Western World
- Greece and Rome
- Romanesque and Gothic
- Renaissance
- Baroque, rococo, neoclassicism, romanticism
- Twentieth century

The Civil War and Reconstruction
- Causes of the war
- Political situations and the war between 1861 and 1865
- Reconstruction

Contemporary Western Europe
- Historical background
- Political systems
- Societal issues
- Integration

Criminal Justice
- Criminal behavior
- Police and criminal justice system
- Court system
- Corrections

Drug and Alcohol Abuse
- Pharmacological principles
- Classification of drugs
- Properties of alcohol, sedative hypnotics, narcotic analgesics, stimulants, hallucinogens and other drugs
- Prevention/treatment

Foundations of Education
- Contemporary issues in education
- Past and current influences on education
- Interrelationships between contemporary issues and influences, past or current, in education

Fundamentals of Counseling
- Counselor roles and functions
- Counseling relationships
- Theoretical approaches
- Social and cultural foundations

General Anthropology
- Anthropology branches, methodologies, and theoretical perspectives
- Physical anthropology and archeology
- Nature of culture
- Social, economic, political, and religious organization

A History of the Vietnam War
- The first Vietnam War (1945-1954)
- LBJ goes to war (1964-1965)
- Tet Offensive (1968)
- Vietnamizing the war (1968-1973)
- Peace is at hand (1968-1973)

Human/Cultural Geography
- Basic facts and concepts
- Culture and environment
- Spatial processes
- Regional geography

Introduction to Law Enforcement
- U.S. criminal justice and police systems
- Issues and organization of police
- Constitutional law and precedents

Introduction to the Modern Middle East
- Geography and early history
- Nineteenth and early twentieth century
- First and Second World wars
- Soviet-U.S. rivalry, Suez Crisis, Iranian Revolution, Persian Gulf Crisis

Lifespan Developmental Psychology
- Biological development
- Perception, learning, and memory
- Social, emotional, and personality development
- Cognition and language

BUSINESS

Business Law II
- Sales of goods
- Debtor and creditor relations
- Business organizations
- Property
- Commercial paper

Business Mathematics
- Basic computation — fractions, percents, ratios, graphs
- Business applications

Introduction to Business
- Economic, international, government, and ownership issues
- Management process
- Human resources and production
- Marketing, finance

Introduction to Computers with Programming in BASIC
- Hardware, software, data management
- Communications and connectivity
- BASIC: I/O, expressions, loops, conditionals, strings, arrays

Management Information Systems
- Systems theory, analysis and design of systems
- Hardware and software
- Database management
- Telecommunications
- Management of the MIS functional area, informational support

Money and Banking
- Role of commercial banks and other intermediaries
- Federal Reserve System
- Macroeconomic activity, U.S. policy, international system

Personal Finance
- Financial goals, budgeting, cash management, economic terminology
- Credit, debt, and major purchases
- Taxes and insurance
- Investments, retirement and estate planning

Personnel/Human Resource Management
- Employment actions
- Training and development
- Performance appraisals
- Compensation and personnel legislation
- Labor relations

Principles of Finance
- Financial statements and ratio analysis
- Capital budgeting
- Time value of money
- Break-even and leverage, common stock

Principles of Financial Accounting
- General concepts and principles
- Financial statements
- Accruals and deferrals
- Current accounts

Principles of Supervision
- Roles and responsibilities
- Management functions
- Legal issues, stress management, unions, productivity and quality concerns

Organizational Behavior
- Individual and group processes and characteristics
- Organizational processes and behavior
- Change and development processes

APPLIED TECHNOLOGY

Technical Writing
- Elements of technical reports
- Style, grammar, mechanics, organizing
- Types of technical reports

HUMANITIES

Ethics in America
- Ethical traditions
- Analysis of issues in relationships
- Tradition versus analysis of issues

Introduction to World Religions
- Dimensions and approaches
- Primal religions, Hinduism, Buddhism, Confucianism, Taoism, Judaism, Christianity, Islam

Principles of Public Speaking
- Topics and purposes
- Organization and supporting materials
- Language, style, and delivery
- Audience analysis and adaptation

PHYSICAL SCIENCE

Astronomy
- The solar system
- The sun and stars
- Our galaxy and the universe
- Measurement

Environment and Humanity: The Race to Save the Planet
- Ecological concepts
- Human impact on the environment
- Environmental management
- Political processes and the future

Here's to Your Health
- Mental health and behavior
- Human development
- Substance use and abuse
- Fitness and nutrition
- Risk factors and disease
- Consumer awareness, environmental concerns

Physical Geology
- Earth materials
- Surface processes
- Internal earth processes
- Applications

Principles of Physical Science I
- Physics: Newton's laws, energy, thermodynamics optics, electricity, and magnetism
- Chemistry: matter, atomic theory and structure, chemical reactions

NOTE:

MOST SCHOOLS DO NOT ACCEPT ALL TESTS FOR ALL PROGRAMS. CHECK WITH THE SCHOOL FROM WHICH YOU EXPECT TO RECEIVE CREDIT.

DANTES

Subject Standardized Tests

Fact Sheet

Study Guide

TECHNICAL WRITING
SF 820

TEST INFORMATION

This test was developed to enable schools to award credit to students for knowledge equivalent to that which is learned by students taking the course. The test contains 100 items. The American Council on Education recommends that credit be awarded for a minimum of 58 correct responses, which is a standard score of 46. This minimum credit-awarding score is equal to the mean score of students in the norming sample who received a grade of C in the course. However, some schools set their own standards for awarding credit and may require a higher score. Students should obtain this information from the institution where they expect to receive credit.

Students may write a short essay on one of the technical topics provided. The essay will not be scored; however, a copy of the essay will be provided with the score report or transcript. Many colleges require an essay. Students, therefore, may wish to contact the schools where scores are to be sent to learn if an essay is required.

CONTENT

The following topics, which are commonly taught in courses on this subject, are covered by this examination.

	Approximate Percent
I. Theory and practice of Technical Writing	10%
A. Establishing goals	
B. Analyzing the audience	
C. Ensuring the validity of data and sources	
II. Purpose, content, and organizational patterns of common types of technical documents	31%
A. Reports	
1. Progress/inspection reports	
2. Feasibility reports	
3. Research/laboratory reports	
B. Correspondence	
1. Memos	
2. Letters	
3. Resumes	
C. Manuals	
1. Instructions	
2. Procedures	
3. Process descriptions	
D. Proposals	
III. Elements of various technical reports	31%
A. Titles	
B. Summaries/Abstracts	
C. Headings	
D. Definitions	
E. Conclusions	
F. Recommendations	
G. Graphics	
H. Report supplements (glossary, footnotes, appendices, indices)	
I. Page Design	

	Approximate Percent
IV. Technical editing	28%
A. Clarity	
B. Completeness	
C. Accessibility	
D. Conciseness	
E. Correctness	
F. Sequence	
G. Unity	
H. Tone	

Questions on the test require candidates to demonstrate the following abilities. Some questions may require more than one of these abilities.

- Knowledge of basic facts and terms (about 35-45 percent of the examination)
- Understanding of concepts and principles (about 30-40 percent of the examination)
- Ability to apply knowledge to specific problems and situations (about 25-35 percent of the examination)

SAMPLE QUESTIONS

1. It is customary to place definitions of terms in all of the following parts of a technical document EXCEPT the

   (A) abstract
   (B) introduction
   (C) footnotes
   (D) glossary

2. Which of the following organizational patterns would be appropriate to use when describing a mechanism?

   I. Order of assembly
   II. Spatial order
   III. Operating order

   (A) II only
   (B) III only
   (C) I and III only
   (D) I, II, and III

3. Which of the following graphics is best for displaying continuous change over time?

   (A) Bar chart
   (B) Line graph
   (C) Schematic diagram
   (D) Table

4. The major difference between proposals and many other technical documents is that proposals are

   (A) long and formal
   (B) written by committees
   (C) overtly persuasive
   (D) presented orally as well as in writing

5. Laboratory reports customarily contain all of the following sections EXCEPT

   (A) materials and methods
   (B) results
   (C) discussions
   (D) recommendations

Questions 6-7 refer to the following group of numbered sentences:

1. The rabbits did not build up as much fat in the walls of their arteries as expected.

2. One group was left unattended except at feeding time, while the other received some old-fashioned tender loving care (TLC) every day for at least one hour.

3. Researchers at Ohio State University made an unexpected finding while feeding a group of rabbits high-cholesterol diets in preparation for an experiment.

4. The TLC rabbits developed only one-third as many fatty deposits as the others.

5. Investigating further, the team fed high doses of cholesterol to two new groups of rabbits.

6. One researcher mentioned that she had given these rabbits special treatment, greeting and cuddling each four to five times a day.

6. If the sentences are arranged into a coherent paragraph, which sentence will come third in the sequence?

(A) 1
(B) 2
(C) 5
(D) 6

7. If the sentences now numbered 1 and 6 were combined into one sentence, the new sentence would begin with which of the following?

(A) When
(B) Although
(C) Because
(D) Thus

8. One significant difference between technical language and lay language is that technical language tends to be more

(A) subjective
(B) sophisticated
(C) abstract
(D) exact

9. Which of the following does NOT need revision to correct an error in parallel structure?

(A) We found it impossible either to replicate their results or even approximate them.
(B) The ideal environment for colonizing plants like ragweed is bare, windswept, and, of course, contains little groundwater.
(C) The platypus is an animal with mammalian, reptilian, and avian aspects and which illustrates the difficulty of exact biological classification.
(D) The proposed pertussis vaccine not only is as effective as the vaccine currently in use but also poses little or no health risk to children under two years of age.

10. Which of the following is a major flaw in a classification?

(A) Overlap in the categories
(B) Use of the same criterion throughout
(C) Division into more than ten groups and subgroups
(D) Categories with different numbers of subdivisions

In each of the following sentences, some part of the sentence or the entire sentence is underlined. Beneath each sentence you will find four ways of phrasing the underlined part. The first choice (A) repeats the original; the other three are different. If you think the original is better than any of the alternatives, choose answer A. Otherwise, choose one of the others. In choosing answers, pay attention to grammatical correctness, appropriate word choice, and smoothness and effectiveness of sentence construction.

11. After heating for twelve minutes, the sample was dried.

    (A) After heating for twelve minutes, the sample
    (B) After heating the sample for twelve minutes, it
    (C) After it had been heated for twelve minutes, the sample
    (D) After twelve minutes of heating the sample, it

12. Ibuprofen is a powerful pain reliever and it has only recently been made available to the public in nonprescription doses.

    (A) Ibuprofen is a powerful pain reliever and has only recently
    (B) Ibuprofen, a powerful pain reliever, has only recently
    (C) Ibuprofen is a powerful analgesic that only has recently
    (D) Only the potent analgesic ibuprofen has recently

REFERENCE PUBLICATIONS

Brusaw, C.T., Alred, G.J., and Oliu, W.E. *Handbook of Technical Writing*. 4th. ed. New York: St. Martin's Press, 1993.

Kennedy, George E. and Montgomery, Tracy T. *Solving Problems Through Technical and Professional Writing*. New York: McGraw-Hill, Inc., 1993.

Lannon, John M. *Technical Writing*. 5th ed. Glenview, IL: Scott, Foresman and Co./Little Brown College Division, 1991.

Markel, Michael H. *Technical Writing: Situations and Strategies*. 3rd ed. New York: St. Martin's Press, 1992.

Mills, Gordon H. and Walter, John A. *Technical Writing*. 5th ed. New York: Holt, Rinehart, and Winston, 1986.

Olsen, Leslie A. and Huckin, Thomas N. *Technical Writing and Professional Communication*. 2nd. ed. New York: McGraw-Hill, 1991.

TECHNICAL WRITING JOURNALS

Journal of Technical Writing and Communications
Technical Communication
The Technical Writing Teacher

CREDIT RECOMMENDATIONS

The Center For Adult Learning and Educational Credentials of the American Council on Education (ACE) has reviewed and evaluated the DANTES examination development process. The American Council on Education has made the following recommendations:

Area or Course Equivalent:	Technical Writing
Level:	Lower level baccalaureate
Amount of Credit:	Three (3) semester hours
Minimum Credit-awarding Score:	Standard score of 46 (58 correct responses)
Source:	ACE Commission on Educational Credit and Credentials

TECHNICAL INFORMATION

Sixty-eight colleges and universities participated in the norming during the fall of 1993. These schools ranged from large state universities to small private schools. Geographic distribution was wide, with schools from all parts of the United States involved. A total of 1,257 students who were completing or had recently completed an elementary course in this subject took part in the norming.

Raw scores were converted to a standard score scale with a mean of 50 and a standard deviation of 10 using the formula $S = AX + B$. The A and B parameters were generated from a mean of 64.0533 and a standard deviation of 13.9045. The parameters are: $A = .7192$ and $B = 3.9329$.

NORMING STATISTICS

(based on 1,255 students)

Number of items	100
Mean (Raw Score)	64.1
Median (Raw Score)	65.1
Standard deviation (Raw Score)	13.9
KR-20 Reliability	.90
Standard Error of measurement (Raw Score)	4.3
(Scaled Score)	3.1

Average Grade/Test Score Correlation, based on a sample of 1,121 students at 61 institutions (an average of within-school correlations, weighted by the number of students tested in each school):	.48
Average scaled score for students whose teachers gave them a grade of A for the course:	56
Average scaled score for students whose teachers gave them a grade of B for the course:	50
Average scaled score for students whose teachers gave them a grade of C for the course:	46

Colleges and universities that would like to have review copies of tests, additional information about the national norming, or assistance in local norming or score validation studies should write to: DANTES Program, Mail Stop 31-X, Educational Testing Service, Princeton, New Jersey 08541-0001.

Correct responses: 1.A; 2.D; 3.B; 4.C; 5.D; 6.D; 7.A; 8.D; 9.D; 10.A; 11.C; 12.B

HOW TO TAKE A TEST

You have studied hard, long, and conscientiously.

With your official admission card in hand, and your heart pounding, you have been admitted to the examination room.

You note that there are several hundred other applicants in the examination room waiting to take the same test.

They all appear to be equally well prepared.

You know that nothing but your best effort will suffice. The "moment of truth" is at hand: you now have to demonstrate objectively, in writing, your knowledge of content and your understanding of subject matter.

You are fighting the most important battle of your life—to pass and/or score high on an examination which will determine your career and provide the economic basis for your livelihood.

What extra, special things should you know and should you do in taking the examination?

BEFORE THE TEST

YOUR PHYSICAL CONDITION IS IMPORTANT

If you are not well, you can't do your best work on tests. If you are half asleep, you can't do your best either. Here are some tips:

1. Get about the same amount of sleep you usually get. Don't stay up all night before the test, either partying or worrying—DON'T DO IT.
2. If you wear glasses, be sure to wear them when you go to take the test. This goes for hearing aids, too.
3. If you have any physical problems that may keep you from doing your best, be sure to tell the person giving the test. If you are sick or in poor health, you really cannot do your best on any test. You can always come back and take the test some other time.

AT THE TEST

EXAMINATION TECHNIQUES

1. Read the *general* instructions carefully. These are usually printed on the first page of the examination booklet. As a rule, these instructions refer to the timing of the examination; the fact that you should not start work until the signal and must stop work at a signal, etc. If there are any *special* instructions, such as a choice of questions to be answered, make sure that you note this instruction carefully.

2. When you are ready to start work on the examination, that is as soon as the signal has been given, read the instructions to each question booklet, underline any key words or phrases, such as *least*, *best*, *outline*, *describe*, and the like. In this way you will tend to answer as requested rather than discover on reviewing your paper that you *listed without describing,* that you selected the *worst* choice rather than the *best* choice, etc.

3. If the examination is of the objective or so-called multiple-choice type, that is, each question will also give a series of possible answers: A, B, C, or D, and you are called upon to select the best answer and write the letter next to that answer on your answer paper, it is advisable to start answering each question in turn. There may be anywhere from 50 to 100 such questions in the three or four hours allotted and you can see how much time would be taken if you read through all the questions before beginning to answer any. Furthermore, if you come across a question or a group of questions which you know would be difficult to answer, it would undoubtedly affect your handling of all the other questions.

4. If the examination is of the essay-type and contains but a few questions, it is a moot point as to whether you should read all the questions before starting to answer any one. Of course if you are given a choice, say five out of seven and the like, then it is essential to read all the questions so you can eliminate the two which are most difficult. If, however, you are asked to answer all the questions, there may be danger in trying to answer the easiest one first because you may find that you will spend too much time on it. The best technique is to answer the first question, then proceed to the second, etc.

5. Time your answers. Before the examination begins, write down the time it started, then add the time allowed for the examination and write down the time it must be completed, then divide the time available somewhat as follows:

    a. If 3 ½ hours are allowed, that would be 210 minutes. If you have 80 objective-type questions, that would be an average of about 2 ½ minutes per question. Allow yourself no more than 2 minutes per question, or a total of 160 minutes, which will permit about 50 minutes to review.

    b. If for the time allotment of 210 minutes, there are 7 essay questions to answer, that would average about 30 minutes a question. Give yourself only 25 minutes per question so that you have about 35 minutes to review.

6. The most important instruction is *to read each question* and make sure you know what is wanted. The second most important instruction is to *time yourself properly* so that you answer every question. The third most important instruction is to *answer every question.* Guess if you have to but include something for each question, Remember that you will receive no credit for a blank and will probably receive some credit if you write something in answer to an essay question. If you guess a letter, say "B" for a multiple-choice question, you may have guessed right. If you leave a blank as the answer to a multiple-choice question, the examiners may respect your feelings but it will not add a point to your score. Some exams may penalize you for wrong answers, so in such cases *only*, you may not want to guess unless you have some basis for your answer.

7. Suggestions

   a. Objective-Type Questions

      (1) Examine the question booklet for proper sequence of pages and questions.

      (2) Read all instructions carefully.

      (3) Skip any question which seems too difficult; return to it after all other questions have been answered.

      (4) Apportion your time properly; do not spend too much time on any single question or group of questions.

      (5) Note and underline key words—*all, most, fewest, least, best, worst, same, opposite.*

      (6) Pay particular attention to negatives.

      (7) Note unusual option, e.g., unduly long, short, complex, different or similar in content to the body of the question.

      (8) Observe the use of "hedging" words—*probably, may, most likely, etc.*

      (9) Make sure that your answer is put next to the same number as the question.

      (10) Do not second guess unless you have good reason to believe the second answer is definitely more correct.

      (11) Cross out original answer if you decide another answer is more accurate; do not erase, *until* you are ready to hand your paper in.

      (12) Answer all questions; guess unless instructed otherwise.

      (13) Leave time for review.

b. Essay-Type Questions

(1) Read each question carefully.

(2) Determine exactly what is wanted. Underline key words or phrases.

(3) Decide on outline or paragraph answer.

(4) Include many different points and elements unless asked to develop any one or two points or elements.

(5) Show impartiality by giving pros and cons unless directed to select one side only.

(6) Make and write down any assumptions you find necessary to answer the question.

(7) Watch your English, grammar, punctuation, choice of words.

(8) Time your answers; don't crowd material.

8. Answering the Essay Question

Most essay questions can be answered by framing the specific response around several key words or ideas. Here are a few such key words or ideas:

M's: manpower, materials, methods, money, management

P's: purpose, program, policy, plan, procedure, practice, problems, pitfalls, personnel, public relations

a. Six basic steps in handling problems:

(1) preliminary plan and background development

(2) collect information, data and facts

(3) analyze and interpret information, data and facts

(4) analyze and develop solutions as well as make recommendations

(5) prepare report and sell recommendations

(6) install recommendations and follow up effectiveness

b. Pitfalls to Avoid

(1) *Taking Things for Granted*
A statement of the situation does not necessarily imply that each of the elements is necessarily true; for example, a complaint may be invalid and biased so that all that can be taken for granted is that a complaint has been registered

(2) *Considering only one side of a situation*
Wherever possible, indicate several alternatives and then point out the reasons you selected the best one.

(3) *Failing to indicate follow up*
Whenever your answer indicates action on your part, make certain that you will take proper follow-up action to see how successful your recommendations, procedures, or actions turn out to be.

(4) *Taking too long in answering any single question*
Remember to time your answers properly.

EXAMINATION SECTION

EXAMINATION SECTION
TEST 1

DIRECTIONS: Each question or incomplete statement is followed by several suggested answers or completions. Select the one that BEST answers the question or completes the statement. *PRINT THE LETTER OF THE CORRECT ANSWER IN THE SPACE AT THE RIGHT.*

1. In the front matter of a typical technical document, which of the following elements is most likely to appear FIRST? 1.____
   A. Title page B. Copyright page
   C. Half title page D. Acknowledgements

2. In preparing a pictograph, it is important to remember that this type of illustration is really a modified 2.____
   A. pie graph B. line graph
   C. flowchart D. bar graph

3. As a general rule, unless a document is extremely technical, it is rarely advisable to use more than _____ main levels of heads. 3.____
   A. 2 B. 3 C. 5 D. 7

4. The division-and-classification method of organizing a document would probably be LEAST useful for 4.____
   A. descriptions of a physical object, such as an engine
   B. the examination of an idea such as a new collective bargaining agreement
   C. persuading management to adopt a solution, such as an improved etching process, over its alternatives
   D. giving instructions, such as the steps necessary for preparing two surfaces for gluing

5. Which of the following sentences is written in the subjunctive mood? 5.____
   A. Take these books with you when you go.
   B. How am I to play without skates?
   C. Tonight the moon is full.
   D. The food would spoil if you were to turn off the refrigerator.

6. To show the proper sequence in which the parts of an object fit together, the BEST illustrative option is probably the 6.____
   A. exploded-view diagram B. schematic diagram
   C. cutaway drawing D. cross-section

7. In Standard Generalized Markup Language (SGML), the end of an element is signified by the symbol 7.____
   A. < B. > C. / D. Ω

8. The primary purpose of an abstract is to 8.___
A. consolidate the principal points of a report in one place
B. enable readers to determine whether to read the work in full
C. give readers enough background on the topic so that they will understand the detailed information in the body of the document
D. focus the reader's attention and stimulate interest in the document

9. If copyright registration is made before or within _____ of the publication of a document, it is prima facie evidence in court that the copyright is valid. 9.___
A. 6 months B. 1 year C. 5 years D. 30 years

10. In general, the MOST effective sales proposals are the ones that are written to sell 10.___
A. conveniences
B. solutions to problems
C. specific and measurable services
D. concrete goods or equipment

11. Which of the following is the International System's symbol for units of acceleration? 11.___
A. N/m^2 B. m/s^2 C. Ω D. m/s

12. In the copyediting of a document, which of the following symbols is a general indicator that is used to mark the position of an error? 12.___
A. Caret B. Parenthesis
C. Ellipse D. Brackets

13. The MAIN difference between a progress report and an activity report is that 13.___
A. progress reports provide budget-related information
B. activity reports are submitted on a fixed schedule
C. activity reports provide information about problems that have been encountered
D. progress reports are submitted by a vendor to a client company

14. In most business applications, a technical document should be written as if the primary readers are to be 14.___
A. line workers
B. other technical professionals
C. lay persons
D. higher-level decision makers

Questions 15-16.

DIRECTIONS: Questions 15 and 16 refer to the following group of numbered sentences.

I. One of the most easily manipulated substances on earth, glass can be made into ceramic tiles that are composed of over 90% air.

II. NASA's space shuttles are the first spacecraft ever designed to leave and re-enter the earth's atmosphere while remaining intact.

III. These ceramic tiles are such effective insulators that when a tile emerges from the oven in which it was fired, it can be held safely in a person's hand by the edges while its interior still glows at a temperature well over 2000°F.

IV. Eventually, the engineers were led to a material that is as old as our most ancient civilizations -- glass.

V. Because the temperature during atmospheric re-entry is so incredibly hot, it took NASA's engineers some time to find a substance capable of protecting the shuttles.

15. If the sentences are arranged into a coherent paragraph, which sentence would come second in the paragraph? 15.___
A. I B. II C. IV D. V

16. Assume that the sentences above have been rearranged into a coherent paragraph. Which of the following devices would be most effective in combining the first two sentences? 16.___
A. Making the first sentence into a dependent clause beginning with *While*
B. Making the first sentence into a dependent clause beginning with *Because*
C. Using a comma and the coordinating conjunction *but*
D. Using a comma and the coordinating conjunction *and*

17. A potential disadvantage associated with using the decreasing-order-of-importance method of organizing a report is that it 17.___
A. does little to ensure that a hurried reader will absorb the most important points
B. does not make a strong initial impression
C. may not encourage a comprehensive examination of the issue
D. may not immediately attempt to get the reader's attention

18. Which of the following symbols is referred to as an ampersand? 18.___
A. @ B. AE C. & D. ∞

19. Which of the following items is not usually included on the title page of a formal report? 19.___
A. The date or dates of the report
B. The name of the writer, principal investigator, or compiler

C. A list of all the major headings in the report
D. The name of the organization for which the writer works

20. <title>Third Quarter Results</title> 20.____
The item above is an example of a
A. title tagged in Hypertext Markup Language (HTML)
B. title tagged in Standard Generalized Markup Language (SGML)
C. Formatting Output Specification Instance (FOSI)
D. Document Type Definition (DTD) for a specific document

21. Which of the following steps in preparing a document is typically performed FIRST? 21.____
A. Gathering information
B. Analyzing the audience
C. Writing a first draft
D. Organizing the document

22. In a document with a series of related equations appearing on the same page, the equations should be 22.____
A. justified left
B. justified right
C. aligned on the equal sign
D. centered

23. Which of the following structural illustrations is not representational? 23.____
A. Map
B. Schematic diagram
C. Flowchart
D. Schedule chart

24. Which of the following statements, comparing *trade libel* or *disparagement of goods* to personal libel, is true? 24.____
A. The majority of personal libel lawsuits are brought against individuals, rather than publications.
B. The *right of fair comment* does not apply to personal libel cases.
C. Privileged statements are not a defense in trade libel.
D. The burden of proof is greater in trade libel.

25. In a feasibility report, the purpose of a conclusion is to 25.____
A. summarize the evaluation of each alternative
B. state the alternative that best meets the criteria
C. present a detailed evaluation of all alternatives under consideration
D. describe the problems that led to the report

KEY (CORRECT ANSWERS)

1. C	6. A	11. B	16. C	21. B
2. D	7. C	12. A	17. C	22. C
3. B	8. B	13. D	18. C	23. B
4. C	9. C	14. D	19. C	24. D
5. D	10. B	15. D	20. B	25. A

TEST 2

DIRECTIONS: Each question or incomplete statement is followed by several suggested answers or completions. Select the one that BEST answers the question or completes the statement. *PRINT THE LETTER OF THE CORRECT ANSWER IN THE SPACE AT THE RIGHT.*

1. Which of the following is NOT a difference between copyediting and proofreading? 1.____
   A. Copy editors mostly verify changes, rather than make them themselves.
   B. Copyediting occurs earlier in production.
   C. Proofreading marks are mostly marginal rather than interlinear.
   D. Proofreaders compare versions of a document with earlier versions.

2. For documents that contain fairly solid prose set in type, with few or no illustrations, the most commonly used format is 2.____
   A. maximum block, double-spaced
   B. two-column
   C. three-column
   D. single column, aligned right

3. What action is being recommended by the copymarking symbol shown at the right? 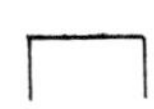 3.____
   A. Aligning text vertically
   B. Closing up a space
   C. Equalizing space
   D. Moving text up

4. Which of the following is the symbol for the International System's basic unit of measurement for electric current? 4.____
   A. Ω B. W C. Hz D. A

5. Typically, which of the following would be checked LAST in the copyediting process? 5.____
   A. Illustrations B. Margins
   C. Grammar D. Consistency

6. Which of the following sentences represents the BEST stylistic choice? 6.____
   A. There must be a thorough review of the tables by engineering personnel.
   B. The tables must be reviewed thoroughly by engineering personnel.
   C. Review of the tables by engineering personnel must be thorough.
   D. The engineering personnel must review the tables thoroughly.

7. A rule of thumb for executive summaries is that they should be about ______% of the length of the report or document they precede. 7.___
   A. 5 B. 10 C. 20 D. 30

8. When writing a proposal, the decision whether to include all the alternatives to a proposed solution will be influenced most strongly by 8.___
   A. the expected length of the proposal
   B. the writer's knowledge of existing bias
   C. financial restrictions
   D. the resources immediately available for implementation

9. Which of the following types of information should typically be EXCLUDED from an abstract? 9.___
   A. The subject of the study
   B. A background of the study
   C. The results obtained
   D. The scope of the study

10. Which of the following is an operational sign? 10.___
   A. = B. ⩽ C. ÷ D. >

11. Which of the following is a general rule for writing correspondence? 11.___
   A. The inside address should be centered.
   B. Address women without a professional title as *Miss*.
   C. The left margin should be at least one inch.
   D. Use *To Whom It May Concern* if the name or title of the recipient is unknown.

12. An important guideline for writing in most documents is that the main idea of a sentence should appear in the structural core of the sentence, the subject and verb of the independent clause. Which of the following sentences best demonstrates this? 12.___
   A. In the twentieth century, the rise in the cancer death rate has paralleled the advances in technology.
   B. Management's policies concerning standards of behavior must be firmly maintained.
   C. It is possible to apply for financial assistance by completing either of two printed forms.
   D. The last decade has seen a remarkable increase in the practice known as telecommuting.

13. In the opening paragraph of a process description, it is most customary for the writer to 13.___
   A. explain the relative difficulty or ease of the process
   B. tell the reader why it is important to become familiar with the process
   C. supply an anecdote or quotation that will stimulate the reader's interest
   D. give a brief overview of the process

14. The MAIN problem with using the ascending arrangement of information within a technical document is that it 14.____
 A. employs a flawed rationale for emphasis
 B. makes supporting details unclear without a main message up front
 C. cannot be used in manuals
 D. forces readers to leap to conclusions

15. In certain situations, it is sometimes preferable to use the passive rather than the active voice of a verb. It would NOT be a reason to use the passive voice in a sentence in order to 15.____
 A. add energy to the sentence
 B. create cohesion and focus
 C. eliminate the naming of an agent that is insignificant or understood
 D. meet readers' expectations in certain scientific documents

16. The body of a proposal should typically include each of the following EXCEPT a(n) 16.____
 A. list of the benefits that will be gained by accepting the proposal
 B. explanation of how a job will be done
 C. cost breakdown for the entire job
 D. description of the methods that will be used to do the job

17. In reviewing a copy editor's marks on a proof, a writer discovers the notation below: 17.____

STORY

Which of the following responses is correct?
 A. Capitalizing only the first letter: Story
 B. Writing the word in italics: *STORY*
 C. Making the entire word lower case: story
 D. Deleting the entire word

18. Under United States copyright law, copyright protection starts from the time the work is 18.____
 A. conceived
 B. created
 C. registered
 D. published

19. When including mathematical material such as formulas or equations in a technical document, the preferred order for signs of aggregation (from first to last used) is 19.____
 A. parentheses, braces, and square brackets: [{()}]
 B. square brackets, braces, and parentheses: ({[]})
 C. braces, parentheses, and square brackets: [({})]
 D. parentheses, square brackets, and braces: {[()]}

20. In most technical documents or reports, the maximum length of any abstract should be about _____ words. 20.____
 A. 100-150
 B. 200-250
 C. 500-750
 D. 750-1000

21. MLA provides each of the following as a rule for using the parenthetical method of documenting sources EXCEPT 21.___
   A. if the author's name is included in the text, include only the title in the citation
   B. when referring to an entire work rather than to a particular page, mention the author's name in the text and omit the parenthetical citation
   C. when documenting sources in text, include only the author and page number in parentheses
   D. when placing parenthetical citations in text, insert them between the closing quotation mark or the last word of the sentence and the end punctuation

22. Which of the following is a descriptive, rather than procedural, markup method? 22.___
   A. HTML
   B. SGML
   C. Templates and styles
   D. Conventional markup

23. Editing a document for accuracy would include 23.___
   A. correcting words for spelling, punctuation, capitalization, and hyphenation
   B. visually measuring segments of graphs to determine whether they are proportional to their numeric values
   C. using space to distinguish rows and columns
   D. checking whether hypertext links are functioning properly

24. In writing and editing a document for translation into a foreign language, each of the following is an important guideline to follow EXCEPT 24.___
   A. avoiding humor and wordplay
   B. using short sentences
   C. using modifying phrases to make descriptions more specific
   D. avoiding acronyms and abbreviations

25. Brief explanations beneath or above an illustration in a document are referred to as 25.___
   A. nomographs
   B. legends
   C. labels
   D. captions

KEY (CORRECT ANSWERS)

1. A	6. D	11. C	16. A	21. A
2. B	7. B	12. A	17. A	22. B
3. D	8. B	13. B	18. B	23. B
4. D	9. B	14. B	19. D	24. C
5. B	10. C	15. A	20. B	25. D

EXAMINATION SECTION
TEST 1

DIRECTIONS: Each question or incomplete statement is followed by several suggested answers or completions. Select the one that BEST answers the question or completes the statement. *PRINT THE LETTER OF THE CORRECT ANSWER IN THE SPACE AT THE RIGHT.*

1. Each of the following is a guideline or rule of thumb for the use of lists in technical documents EXCEPT 1.____
   A. making items of equal or comparable length
   B. using only words, phrases, or short sentences
   C. using parallel structure throughout
   D. when rank or sequence is important, using bullets

2. The most important difference between a *purpose statement* and a *message statement* in setting goals for a document is that a 2.____
   A. message statement is most likely to be directed at a document's secondary readers
   B. purpose statement describes what the reader will learn if he or she continues to read
   C. purpose statement is a crystallization of the document's main idea
   D. message statement is only used in documents whose purpose is to persuade

3. Most often, the internal proposal adopts a format that is comparable or equivalent to the 3.____
   A. formal report
   B. memo
   C. sales proposal
   D. formal letter

4. In technical documents, the left-hand vertical column in tabular illustrations is sometimes referred to as the 4.____
   A. horizon rule
   B. dummy
   C. boxhead
   D. stub

5. The typesetter in document preparation is sometimes referred to as a(n) 5.____
   A. compositor
   B. iterator
   C. justifier
   D. stringer

6. An irregular right margin in a document is said to be 6.____
   A. aliased
   B. ragged
   C. unregistered
   D. justified

7. In page design, the term *widows and orphans* refers to 7.____
   A. mock-ups of how a finished publication will look
   B. pages appearing at the end of a chapter or section that have two lines (or fewer) of type
   C. the accidental hierarchical patterns that are sometimes applied when using captions to summarize passages
   D. words, or word segments, that are stranded at the end of a column or carried over to the next page

8. <u>Loch Lomond</u> 8.___
The above copymarking symbol, appearing in the corrected proofs of a document, is an indication that in revising the author should
A. delete the selection
B. underline the selection
C. substitute italic type for Roman
D. type the word in all caps

9. Usually, memos should begin by stating the main point. 9.___
The exception(s) to this rule is/are when the
I. audience is captive
II. reader is likely to be highly skeptical
III. writer is in disagreement with persons in positions of higher authority

The CORRECT answer is:
A. I *only*
B. I, II
C. II, III
D. All of the above

Questions 10-11.

DIRECTIONS: Questions 10 and 11 refer to the following paragraph.

The earliest writing materials were used in Mesopotamia more than 5,000 years ago. They were thin clay tablets. Although the tablets were cheap and easy to produce, they had two major disadvantages. They were difficult to store, and once the clay had dried and hardened a person could not write on them.

10. Which of these is the BEST way to combine the first two sentences? 10.___
A. The earliest writing materials were thin clay tablets, used in Mesopotamia more than 5,000 years ago.
B. The earliest writing materials were used in Mesopotamia more than 5,000 years ago and were thin clay tablets.
C. Thin clay tablets, the earliest writing materials were used in Mesopotamia more than 5,000 years ago.
D. The earliest thin clay tablets were used as writing materials in Mesopotamia more than 5,000 years ago.

11. Which of these sentences could BEST be added at the end of this paragraph? 11.___
A. Mesopotamia was a civilization in the Middle East, near Egypt.
B. The Egyptians used to write on the bark of the papyrus plant, a thin reed that grew all over the delta of the Nile.
C. Mesopotamian people are also remembered for their stone artwork.
D. Unfortunately, it wasn't until many years later that people figured out more convenient ways to write.

12. A column graph is usually better than a bar graph for comparing 12.____
   A. groups of items on a single point
   B. speed
   C. trends over time
   D. distance

13. Customer- and client-orientation documents tend to use an organizational framework driven by 13.____
   A. numerical or alphabetical order
   B. chronology
   C. description
   D. geography

14. Cases of *trade libel* associated with the writing of technical information usually involve 14.____
   A. advertising copy
   B. abstracts
   C. private correspondence or memos
   D. manuals

15. What action is being recommended by the copy-marking symbol shown at the right? 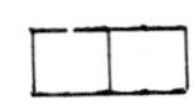 15.____
   A. Transposing adjacent letters
   B. Indenting two ems
   C. Inserting a graphic or chart
   D. Carrying text back to the preceding line

16. Which of the following words contains a diacritical mark known as a dieresis? 16.____
   A. Cliché B. Crêpe
   C. Garçon D. Coöperation

17. Preparatory outlines for technical documents are typically used for each of the following purposes EXCEPT 17.____
   A. maintaining continuity through interruptions
   B. setting schedules and deadlines
   C. determining the audience's background
   D. discovering missing or incomplete information

18. In most applications, a progress report is written in the format of a 18.____
   A. letter B. list C. memo D. table

19. In which of the following sentences is a subject complement underlined? 19.____
   A. Our webmaster has created <u>a new database</u>.
   B. The machine has been <u>updated</u>.
   C. The <u>older</u> models in the product line need to be phased out.
   D. The new technology makes our product <u>obsolete</u>.

20. Of the following types of documents, which usually represents the latest stage in the production process? 20.___
A. Page proof
B. Typescript
C. Galley
D. Blueline

21. Which of the following is a commonly used guideline for writing letter headings? 21.___
A. The date usually goes beneath the last line of the address.
B. Words like Street, Avenue, First, or West should be abbreviated rather than spelled out.
C. In the date, the name of the month is usually abbreviated.
D. The writer's name should appear after the date.

22. In document preparation, SGML is most useful for 22.___
A. formatting
B. text editing
C. production and information management
D. providing visual flexibility

23. Which of the following types of technical documents is most likely to include both an abstract and executive summary? 23.___
A. Sales proposal
B. Laboratory report
C. Process descriptions
D. Formal report

24. Which of the following errors is most commonly committed in the writing of mathematical text? 24.___
A. Misuse of symbols
B. Dangling participle
C. Incomplete sentences
D. Misspelled *jargon* words

25. In typography, *point size* is measured as 25.___
A. the distance from the bottom of descenders to the top of the x-height
B. the distance from the top of ascenders to the bottom of the x-height
C. the distance from the bottom of descenders to the top of ascenders
D. equal to the x-height

KEY (CORRECT ANSWERS)

1. D	6. B	11. D	16. D	21. A
2. B	7. D	12. C	17. C	22. C
3. B	8. C	13. D	18. A	23. D
4. D	9. C	14. A	19. B	24. B
5. A	10. A	15. B	20. D	25. C

TEST 2

DIRECTIONS: Each question or incomplete statement is followed by several suggested answers or completions. Select the one that BEST answers the question or completes the statement. *PRINT THE LETTER OF THE CORRECT ANSWER IN THE SPACE AT THE RIGHT.*

1. Which of the following is a guideline for the use of tables as illustrations in a technical document? 1.____
   A. Points of comparison should be listed across the top of the table.
   B. Vertical and horizontal lines should be used whenever possible to separate rows and columns.
   C. Columns of numbers should be aligned to the right.
   D. All tables should be referred to as *figures*.

2. In technical documents, the "&" symbol should be used only 2.____
   A. if it appears in an official company name
   B. if space is limited
   C. in scholarly abbreviations
   D. in mathematical material

3. If the purpose of a document is to convince an organization to fund a project, which of the following organizational frameworks is probably most appropriate for the document? 3.____
   A. Cause-and-effect
   B. Comparison
   C. Spatial
   D. Order-of-importance

4. Other than a proposal, the documents whose purpose is most likely to include persuasion is the 4.____
   A. memorandum
   B. formal report
   C. activity report
   D. procedural manual

5. In writing about statistical relationships, which of the following symbols should be used to indicate a mean value? 5.____
   A. μ
   B. Δ
   C. §
   D. *f*

6. The copymarking symbol shown at the right, ② appearing in the corrected proofs of a document, is an indication that in revising, the author should 6.____
   A. write out the number as *two*
   B. delete this number
   C. use a Roman numeral or other method for sequencing
   D. indent

Questions 7-9.

DIRECTIONS: Questions 7 through 9 refer to the paragraph below.

Pheromones are chemical compounds. These compounds contain messages that influence animal behavior. The activity of an entire bee colony is governed mostly by pheromones that are passed from one bee to another. Many of the most important pheromones are released by the queen bee. Food gathering and storage, raising young, and reproduction in a bee colony are all controlled by pheromones.

7. Which of these sentences would NOT belong in this paragraph? 7.___
   A. The pheromones are spread out through the air when a bee fans its wings.
   B. Sometimes the pheromones are passed mouth-to-mouth between bees.
   C. There are many different kinds of bees in the world.
   D. A bee receives an order from the queen through a pheromone message.

8. How can the first two sentences in this paragraph BEST be combined? 8.___
   Pheromones are chemical compounds,
   A. which compounds contain messages that influence animal behavior
   B. and contain messages that influence animal behavior
   C. and these compounds contain messages that are influential in the behavior of animals
   D. but they also contain messages that influence animal behavior

9. Which of these sentences could BEST be added after the last sentence in this paragraph? 9.___
   A. Without these pheromones, the colony would become disorganized and die out.
   B. Other animals use pheromones to send signals.
   C. Pheromones are used in perfumes and colognes as a way to attract people.
   D. Even simple plants like algae use pheromones.

10. Which of the following must ALWAYS be done before the writing of a document can begin? 10.___
   I. Organizing the information
   II. Devising a purpose
   III. Identifying the audience
   IV. Developing an introduction

   The CORRECT answer is:
   A. I, III B. II, III C. III, IV D. II, IV

11. A change in the form of a word to show a specific meaning or a grammatical relationship to another word is known as 11.___
   A. mutation B. inflection
   C. conjugation D. declension

12. Which of the following sentences contains the strongest verb? 12.____
    A. The report will cover changes made in the final phase of the project.
    B. A decision has finally been reached by the board.
    C. If Council members agree about the park's needs, they will hire a consulting firm.
    D. The Chamber of Commerce will hold a meeting for discussion of the expansion question.

13. Which of the following statements, concerning abstracts, summaries, and introductions, is true? 13.____
    A. An executive summary should generally run no longer than 200-300 words.
    B. An introduction is intended to be a snapshot of a longer work.
    C. An abstract never varies in the type of information it provides.
    D. An executive summary may be the only section of a longer work read by some readers, so it must cover the information in the greatest detail.

14. In document production, the right-hand page of a book, numbered with an odd number, is referred to as the 14.____
    A. recto B. intaglio C. verso D. diazo

15. The first element to be presented in a typical informative abstract is typically the 15.____
    A. recommendations
    B. methods used to arrive at the findings
    C. purpose and scope of the document
    D. general conclusions

16. Unlike styles and templates, HTML 16.____
    A. is a descriptive method
    B. does not prescribe the exact look of coded document
    C. is a procedural method
    D. embeds visual consistency

17. The full block style format is suitable *only* for 17.____
    A. memoranda and other brief documents
    B. electronic messages
    C. business letterhead stationery
    D. long, formal reports

18. The first step in editing an online document would typically be to 18.____
    A. run computer checks
    B. edit the illustrations and other nonprose text
    C. survey the document
    D. edit paragraphs and headings for correctness, consistency, and accuracy

19. Each of the following is a useful guideline for creating and using drawings in technical documents EXCEPT 19.____
A. showing equipment from the same point of view at all times
B. always showing a subsystem's relationship to the larger system of which it is part
C. arranging sequenced drawings from left to right or top to bottom
D. if possible, placing labels on the parts themselves

20. Which of the following is the International System's basic unit of measurement for luminous intensity? 20.____
A. Watt
B. Candela
C. Lumina
D. Foot-candle

21. Most *adjustment* letters begin with 21.____
A. an explanation of what caused the problem
B. an explanation of why the problem will or will not be remedied
C. an apology
D. whatever the reader will consider to be good news

22. When a work or part of a work is published under the authority of the copyright owner, a notice of copyright should be placed on all publicly distributed copies. Which of the following is NOT a required element of this notice? 22.____
A. The name of the copyright year
B. The year of the first publication of the work
C. Year of the most recent publication of the work
D. The symbol ©, or the word *Copyright*, or the abbreviation *Copr*.

23. Though usually unnumbered, the title page of a formal report is considered to be page 23.____
A. i
B. ii
C. iii
D. 1

24. In the body of a typical technical document, which of the following elements is most likely to appear LAST? 24.____
A. List of references
B. Chapters
C. Abstract
D. Acknowledgements

25. For illustrating a single change over time, temperature, or other independent variable, the best illustration to use is probably the 25.____
A. line graph
B. bar graph
C. pictograph
D. column graph

KEY (CORRECT ANSWERS)

1. A	6. A	11. B	16. B	21. D
2. A	7. C	12. C	17. C	22. C
3. D	8. B	13. D	18. C	23. A
4. A	9. A	14. A	19. A	24. A
5. A	10. B	15. C	20. B	25. A

EXAMINATION SECTION
TEST 1

DIRECTIONS: Each question or incomplete statement is followed by several suggested answers or completions. Select the one that BEST answers the question or completes the statement. *PRINT THE LETTER OF THE CORRECT ANSWER IN THE SPACE AT THE RIGHT.*

Questions 1-15.

DIRECTIONS: In Questions 1 through 15, some part of the sentence or the entire sentence is underlined. Beneath each sentence you will find four ways of phrasing the underlined part. The first choice (A) repeats the original; the other three are different. If you think the original is better than any of the alternatives, choose answer A. Otherwise, choose one of the others. In choosing answers, pay attention to grammatical correctness, appropriate word choice, and effectiveness of sentence construction.

1. <u>The plastic rivets used in assembling toys have replaced the machine screws</u>, which are easier for workers to handle. 1.____
   A. The plastic rivets used in assembling toys have replaced the machine screws
   B. The plastic rivets, which have replaced the machine screws, are used in assembling toys
   C. The machine screws have since been replaced by the plastic rivets used in assembling toys
   D. Used in assembling toys, the plastic rivets have replaced the machine screws

2. <u>To get accurate results you must sample from several sites, make sure depth measurements are identical, and using the same instruments helps too.</u> 2.____
   A. To get accurate results you must sample from several sites, make sure depth measurements are identical, and using the same instruments helps too.
   B. To get accurate results you must sample from several sites, making sure depth measurements are identical, and to use the same instruments helps too.
   C. To get accurate results you must sample from several sites; make sure depth measurements are identical and using the same instruments helps too.
   D. To get accurate results you must sample from several sites, make sure depth measurements are identical, and use the same instruments.

3. <u>Without proper calibration, a good scientist never can record the correct data.</u> 3.____
   A. Without proper calibration, a good scientist never can record the correct data.
   B. Without proper calibration, the correct results never can be recorded by a good scientist.

C. The correct results, without proper calibration, never can be recorded by a good scientist.
D. A good scientist never can record the correct results; without proper calibration.

4. The programs work together to develop an event calendar, since their sponsoring companies do not have a common vocabulary for events, they are forced to invent their own. 4.___
A. an event calendar, since their sponsoring companies
B. an event calendar. Since their sponsoring companies
C. an event calendar since their sponsoring companies
D. an event calendar since, their sponsoring companies

5. The first steps in the process are as follows, condensation, titration, and evaporation. 5.___
A. follows, condensation, titration, and evaporation
B. follows: Condensation, titration, and evaporation
C. follows: condensation, titration, and evaporation
D. follow condensation; titration, and evaporation

6. Because of the limits of the pulping process, fewer than ten runs are possible using the same fiber mix. 6.___
A. the pulping process, fewer than ten runs are possible
B. the pulping process, less than ten runs are possible
C. the pulping process, only less than ten runs are possible
D. the pulping process, ten runs are fewer than possible

7. Military officials, however, had a difficult time developing codes that could fool Japanese codebreakers, due to the fact that they had an ability to recognize and decode unfamiliar grammatical structures that was often superior to that of Americans. 7.___
A. due to the fact that they had an ability to recognize and decode unfamiliar grammatical structures that
B. due to their ability to recognize and decode unfamiliar grammatical structures
C. simply because of the fact that they had an ability to recognize and decode unfamiliar grammatical structures that
D. whose ability to recognize and decode unfamiliar grammatical structures

8. The connector couples with the pin adjacent to it to form an armature. 8.___
A. The connector couples with the pin adjacent to it to form an armature.
B. The connector couples with the adjacent pin to form an armature.
C. With the pin adjacent to it, the connector couples to form an armature.
D. To form an armature, the connector couples with the pin adjacent to it.

9. This group of compounds is observed to be among the most volatile in commercial use. 9.___
   A. This group of compounds is
   B. This group of compounds are
   C. These groups of compounds is
   D. Often, this compound group are

10. The output device then commences operation, transposing images for users which are later framed. 10.___
   A. transposing images for users which are later framed
   B. transposing images for framed users
   C. transposing images for users. The images are later framed
   D. transposing images for users, which are later framed

11. Known as the articulating joint, in many of the most basic functions achieved by the mechanism, this element plays a key role during power transfer. 11.___
   A. Known as the articulating joint, in many of the most basic functions achieved by the mechanism, this element plays a key role in transferring power in most applications.
   B. During power transfer this element, known as the articulating joint, plays a key role in many of the most basic functions achieved by the mechanism.
   C. A key role is played, during power transfer, in many of the most basic functions achieved by the mechanism, by the element known as the articulating joint.
   D. Playing a key role in many of the most basic functions achieved by the mechanism, this element is the articulating joint during power transfer.

12. The hammermill chamber, the machine's largest structure requires the greatest skill in disassembly. 12.___
   A. The hammermill chamber, the machine's largest structure of the entire machine requires the greatest skill in disassembly.
   B. The hammermill chamber, the largest structure of the entire machine requires the greatest skill in disassembly.
   C. The hammermill chamber, the machine's largest structure, requires the greatest skill in disassembly.
   D. The machine's largest structure the hammermill chamber requires the greatest skill in disassembling.

13. The data were collected incorrectly and are now skewed but they can be corrected by an adjustment in the original equation. 13.___
   A. The data were collected incorrectly and are now skewed but they can be corrected
   B. The data were collected incorrectly and are now skewed, but they can be corrected
   C. The data was collected incorrectly and is now skewed but it can be corrected
   D. The data was collected incorrectly and is now skewed; but it can be corrected

14. The machinist etches the engraving plate applying an acid solution. 14.___
    A. The machinist etches the engraving plate applying an acid solution.
    B. The engraving plate is etched by the machinist applying an acid solution.
    C. Applying an acid solution, the machinist etches the engraving plate.
    D. Etching the engraving plate the machinist applies an acid solution.

15. "Eskimo" entered the English language from the French term, "esquimaux," by which other North American natives used to refer to the natives of the far north. 15.___
    A. the French term, "esquimaux," by which other North American natives used to refer to the natives of the far north
    B. the French term, "esquimaux," while other North American natives used to refer to the natives of the far north
    C. the French term, "esquimaux," which other North American natives used to refer to the natives of the far north
    D. the French term, "esquimaux," although other North American natives used to refer to the natives of the far north

16. Which of the following tasks is LEAST likely to be involved in the revision stage of preparing a document? 16.___
    A. Making sentences and paragraphs more coherent
    B. Organizing the information
    C. Selecting and designing visuals
    D. Selecting more precise words

17. Which of the following sentences demonstrates correct usage? 17.___
    A. The number of thefts has increased over the past few months, and so the laboratory doors are now to be locked at all hours.
    B. I am surprised at the amount of errors in this report.
    C. The number of electricity available for the super-collider has been curtailed.
    D. The amount of electrons in the compound has mysteriously changed in the past hour.

18. Editing a document for correctness and consistency would include 18.___
    A. aligning columns of numbers on decimals or imagined decimals
    B. making sure labels, numbers, and titles are present
    C. establishing patterns for placement of callouts and titles
    D. adding columns of numbers to check the totals

19. An introduction to the first progress report in series should include each of the following EXCEPT 19.____
A. a description of the project's current status
B. a definition of all methods and necessary materials
C. an identification of the project by name
D. the date by which the project is to be completed

20. In most technical publications, the purpose of the index is to 20.____
A. list the key terms used in the text and the pages on which they are discussed or referred to
B. provide supplementary text material such as research instruments, tables of data, or troubleshooting guides
C. define terms used in the document
D. list the complete publication data for the materials cited in the document

21. In the back matter of a typical technical document, which of the following elements is most likely to appear LAST? 21.____
A. Glossary B. Index
C. References D. Appendix

22. In documents whose purpose is to persuade, which of the following organizational frameworks is LEAST commonly used? 22.____
A. Chronology
B. Decreasing order of importance
C. Problem to solution
D. Comparison

23. In reviewing the copy editor's marks on a proof, a writer discovers the following notation: (hp) 23.____
Which of the following would provide the desired response?
A. Capitalizing, to read *HP*
B. Underlining, to read hp
C. Writing out the word *horsepower* for the abbreviation *hp*
D. Capitalizing the first word, to read *Hp*

24. In the arena of government proposals, the most important difference between an Invitation to Bid and a Request for Proposals (RFP) is that the 24.____
A. Invitation to Bid includes specifications
B. Invitation to Bid involves testing and evaluation of submitted bids
C. RFP is subject to affirmative action legislation
D. RFP is flexible and open to negotiation

25. Which of the following is an element of mechanical consistency in a technical document? 25.____
A. Documentation B. Screen design
C. Typography D. Syntax

KEY (CORRECT ANSWERS)

1. C
2. D
3. A
4. B
5. C

6. A
7. D
8. B
9. A
10. C

11. B
12. C
13. B
14. C
15. C

16. B
17. A
18. C
19. A
20. A

21. B
22. A
23. C
24. D
25. A

TEST 2

DIRECTIONS: Each question or incomplete statement is followed by several suggested answers or completions. Select the one that BEST answers the question or completes the statement. *PRINT THE LETTER OF THE CORRECT ANSWER IN THE SPACE AT THE RIGHT.*

1. It is generally accepted by writers and managers that the most essential function of technical editing is 1.____
   A. comprehensive editing for style
   B. copyediting for correctness and consistency
   C. comprehensive editing for organization
   D. copyediting for accuracy

2. Which of the following types of documents is most likely to require an introduction? 2.____
   A. Newsletter articles B. Memoranda
   C. Laboratory reports D. Proposals

3. What is the term for the top part of a letter (usually lower case, as in b, d, and h) that rises above the x-height? 3.____
   A. Ascender B. Serif C. Tail D. Kern

4. The MAIN difference between industrial specifications and technical manuals is that 4.____
   A. manuals are written in a more informal tone
   B. specifications contain detailed information that the customer does not need
   C. manuals are written in the imperative mood
   D. specifications are not usually given to the customer

5. Which of the following sentences uses the *active* voice? 5.____
   A. The table was compiled collaboratively by two engineers and a graphic artist.
   B. This procedure for writing and editing documents has been successful for us.
   C. Two additional tables have been left out of the appendix.
   D. The front matter is being printed now, as we speak.

6. In trade libel law, the *puffing doctrine* states that 6.____
   A. an attack on goods sold or manufactured by a plaintiff can be strong enough to impugn the plaintiff's character
   B. private figures, as opposed to public ones, may press libel actions when there is merely negligence and not malice
   C. one can claim the superiority of a good, service, or process as long as no specific assertions of fact are made that are unfavorable to one's rivals
   D. the right of fair comment and criticism is not limited to public officers and candidates for office

7. The primary purpose of an opening to a brief article or report is to 7.___
   A. put the subject in perspective
   B. summarize the topic
   C. give the reader a basis for judging the actual results
   D. stimulate the reader's interest

8. The International System's basic unit of measurement for temperature is the 8.___
   A. degree Celsius B. degree Fahrenheit
   C. newton D. kelvin

9. Which of the following types of visuals is least abstract? 9.___
   A. Cutaway drawing B. Photograph
   C. Exploded drawing D. Line drawing

10. Which of the following is a participial phrase? 10.___
   A. Slowing considerably B. To catch a thief
   C. Stainless steel D. Running water

11. Which of the following is NOT a purpose of the front matter of a formal report? 11.___
   To
   A. give the reader a general idea of the author's purpose in writing the report
   B. list where in the report the reader can find specific chapters, headings, or illustrations
   C. indicate whether the report contains the kind of information the reader is looking for
   D. indicate where additional information about the topic can be found

12. Text that is standard for various documents and can be inserted, with minimal or no revision, in a new document is known as 12.___
   A. verso B. boilerplate
   C. schema D. hard copy

13. A newsletter will typically require more creativity and innovation from a technical writer than other assignments because 13.___
   A. it must be produced at steady periodic intervals
   B. its readership is not a captive audience
   C. it is extremely condensed
   D. it makes greater use of narrative devices

14. Which of the following is not a general principle for the organization of a technical document? 14.___
   A. Arrange information from specific to general
   B. Anticipate a reader's questions
   C. Follow pre-established document structures
   D. Arrange information from new to familiar

15. Which of the following statements is written in the indicative mood? 15.___
   A. If you were to turn your back, he would take advantage of you.
   B. Take your feet off the table, please.
   C. Which one of those belongs to you?
   D. The horses are on the track.

16. The appendix or appendices to a report may contain any of the following EXCEPT 16.___
   A. letters or other supporting documents
   B. case histories
   C. bits of information that the writer was unable to work into the text
   D. computer printouts of raw data

17. When considering the audience for a technical document, the key question typically focuses on 17.___
   A. geography
   B. audience size
   C. prior knowledge and experience
   D. demographic composition

18. Which of the following is a general rule for writing correspondence? 18.___
   A. The heading should be aligned with the right margin.
   B. The body of the letter should begin five spaces below the salutation.
   C. Capitalize each word in the complimentary closing.
   D. The writer's full name should appear four spaces below the complimentary closing.

19. Which of the following is NOT an element of verbal consistency? 19.___
   A. Style B. Syntax C. Spelling D. Semantics

20. Descriptive abstracts are probably LEAST appropriate for 20.___
   A. information surveys
   B. progress reports that combine information from more than one project
   C. conference proceedings
   D. annual reports released to the public

21. What is the term for the process of calculating the length of a document according to the number of characters in a typescript and the specifications for typeface and type size, line length, and page depth? 21.___
   A. Markup B. Imposition
   C. Stitching D. Casting off

22. The FIRST element included in a long sales proposal is typically the 22.___
   A. title page B. advertising literature
   C. transmittal letter D. executive summary

23. The copymarking notation shown at the right, appearing in the corrected proofs of a document, is an indication that in revising, the author should 23.____

replace (wavy underline)

A. underline the word: replace
B. use italics: *replace*
C. use boldface: **replace**
D. delete the word

24. Comprehensive editing, or macro-editing, typically includes each of the following procedures EXCEPT 24.____
A. editing for style
B. addressing legal and ethical issues
C. checking the completeness of parts
D. reshaping sentence structures

25. When using numbers in the text of a technical document, each of the following is a general guideline to follow EXCEPT 25.____
A. given a choice, use the metric system of measurement
B. do not begin a sentence with a figure
C. do not set decimal fractions of less than 1.0 with an initial zero
D. use figures for all quantifiable units of measure, rather than spelling out the numbers

---

KEY (CORRECT ANSWERS)

1. B
2. C
3. A
4. B
5. B

6. C
7. D
8. D
9. B
10. A

11. D
12. B
13. B
14. A
15. D

16. C
17. C
18. D
19. C
20. D

21. D
22. C
23. C
24. C
25. C

EXAMINATION SECTION
TEST 1

DIRECTIONS: Each question or incomplete statement is followed by several suggested answers or completions. Select the one that BEST answers the question or completes the statement. *PRINT THE LETTER OF THE CORRECT ANSWER IN THE SPACE AT THE RIGHT.*

Questions 1-2.

DIRECTIONS: Questions 1 and 2 refer to the following group of numbered sentences.

I. Instead of wheels, the HSST uses two sets of magnets, one which sits on the track, and another that is carried by the train.

II. These magnets generate an identical magnetic field which forces the two sets apart.

III. In the last few decades, railway travel has become less popular throughout the world, because it is much slower than travel by airplane, and not much less expensive.

IV. The HSST's designers say that the train can take passengers from one town to another as quickly as a jet plane -- while consuming less than half the energy.

V. This repellent effect is strong enough to lift the entire train above the trackway, and the train, literally traveling on air, rockets along at speeds of up to 300 miles per hour.

VI. The revolutionary technology of magnetic levitation, currently being tested by Japan's experimental HSST (High Speed Surface Transport), may yet bring passenger trains back from the dead.

1. If the sentences were arranged into a coherent paragraph, which sentence would come third in the sequence? 1.____
   A. I B. III C. IV D. V

2. Which sentence would appear last? 2.____
   A. III B. IV C. V D. VI

3. Which of the following is/are potential disadvantage(s) associated with the sequential or step-by-step method of organizing a technical document? 3.____
   I. It can become monotonous.
   II. It is difficult to follow.
   III. It does not lend itself to emphasis.

   The CORRECT answer is:
   A. I *only* B. I, III C. II, III D. I, II

4. Templates and styles increase the efficiency of preparing documents for distribution because they 4.___
   A. eliminate the step of hard copy markup for design after the text is created
   B. help create a searchable database for the document
   C. weed out errors in consistency and accuracy
   D. specify tags for more than 200 elements of text and other identifiers

5. A verbal identifier, such as the name of a part, in an illustrated diagram is known as a 5.___
   A. callout B. heading C. legend D. caption

6. A descriptive abstract typically includes each of the following EXCEPT information about the 6.___
   A. document's purpose
   B. general conclusions of the document
   C. scope of the document
   D. methods used to arrive at the findings in the document

7. Which of the following is NOT an exclusive right of a copyright owner? 7.___
   To
   A. prepare derivative works based upon the copyrighted work
   B. approve or disapprove of the use of any part of the copyrighted work, attributed or not, in another publication
   C. publicly display the copyrighted work
   D. distribute copies of the work to the public by sale or other transfer of ownership

8. For a report with a number of sections or chapters, the reference section should appear in the 8.___
   A. front matter, directly after the table of contents
   B. back matter, directly after the appendix or appendices
   C. body, at the end of each major section or chapter
   D. body, directly after the end of the report

9. Which of the following words is a nominalization? 9.___
   A. Grunt B. Readership
   C. Substitute D. Consideration

Questions 10-12.

DIRECTIONS: Questions 10 through 12 refer to the following paragraph.

In the early 1970's, the American grizzly bear almost disappeared from Yellowstone National Park after officials closed the park's garbage dumps. The bears had become so dependent on garbage for food that they almost died out. The bears have begun to relearn their old hunting and food-gathering techniques, but there has been an increase in the population.

10. Which of these choices would be the BEST option for writing the last sentence more clearly? 10.___
   A. The bears have begun to relearn their old hunting and food-gathering techniques, however, there has been an increase in the population.
   B. The bears have begun to relearn their old hunting and food-gathering techniques, and there has been an increase in the population.
   C. The bears have begun to relearn their old hunting and food-gathering techniques, and nevertheless there has been an increase in the population.
   D. The bears have begun to relearn their old hunting and food-gathering techniques, there has been an increase in the population.

11. Which of these sentences COULD be effectively added to this paragraph? 11.___
   A. The grizzlies still number fewer than 1,000.
   B. The grizzly has been in decline ever since the Lewis and Clark expedition.
   C. The Yellowstone elk population is in danger also.
   D. The Alaskan grizzly, however, is still going strong.

12. Which of these options demonstrates the BEST way of moving from one idea to another in the last sentence? 12.___
   A. For example, the bears have begun to relearn their old hunting and food-gathering techniques,
   B. In conclusion, the bears have begun to relearn their old hunting and food-gathering techniques,
   C. On the contrary, the bears have begun to relearn their old hunting and food-gathering techniques,
   D. Recently, the bears have begun to relearn their old hunting and food-gathering techniques,

13. Versions of typescript as it moves through various editorial passes are called 13.___
   A. galleys B. iterations
   C. markups D. loops

14. In reviewing the copy editor's marks on a proof, a writer encounters the following notation: ibm . 14.___
   How should the selection be rewritten?
   A. IBM
   B. I.B.M.
   C. International Business Machines
   D. IBM (small caps)

15. Typically, which of the following would be checked FIRST in the copyediting process? 15.___
   A. Type styles B. Mathematical material
   C. Consistency D. Headings

16. Investigative reports most frequently employ the _____ framework in their organization. 16.___
A. chronological B. ascending arrangement
C. cause to effect D. problem to solution

17. For traditionally typewritten material such as reports, which of the following formats is most widely used? 17.___
A. Wide column aligned left
B. Wide column aligned right
C. Traditional two-column
D. Maximum block, double-spaced

18. A version of a document during production that has been superseded by a later version is known as 18.___
A. dead copy B. a proof C. crop D. a galley

19. Which of the following is LEAST likely to be organized chronologically? 19.___
A. A trip report B. An annual report
C. Minutes of meetings D. A laboratory report

20. In writing about statistical relationships, which of the following symbols should be used to indicate probability? 20.___
A. r B. N C. ϕ D. df

21. Editing a document for completeness would include 21.___
A. checking that all illustrations mentioned in the text are present and that the numbers go in sequence
B. comparing line drawings to the actual product
C. comparing references in the text with the illustration
D. checking the title of an illustration for a match with its contents

22. In writing reports, the MAIN difference between introductions and abstracts or executive summaries is that introductions 22.___
A. typically involve the greatest level of detail
B. highlight the major points of the report
C. provide background information for understanding the details in the longer work
D. include information about the purpose, scope, and methodology of the longer work

23. Which of the following is an element of visual consistency in a technical document? 23.___
A. Cross-references B. Style
C. Lists D. Layout

24. Typically, the appendix or appendices of a report appear in the back matter, 24.___
A. at the very beginning
B. directly after the bibliography
C. directly after the glossary
D. directly after the index

25. Which of the following is NOT generally true of hypertext documents? 25.___
They
A. require greater contrast between text and background
B. require less visual consistency than print documents
C. should use navigation aids
D. should use an absolute minimum of italic type

KEY (CORRECT ANSWERS)

1. A
2. B
3. B
4. A
5. A

6. B
7. B
8. C
9. D
10. B

11. A
12. D
13. B
14. A
15. C

16. A
17. D
18. A
19. B
20. C

21. A
22. C
23. D
24. B
25. B

TEST 2

DIRECTIONS: Each question or incomplete statement is followed by several suggested answers or completions. Select the one that BEST answers the question or completes the statement. *PRINT THE LETTER OF THE CORRECT ANSWER IN THE SPACE AT THE RIGHT.*

1. Which of the following sentences is written in *telegraphic* style? 1.____
   A. You should use your right hand to operate the cutoff switch, since the switch will be located on your right side as you face the machine.
   B. One must always be mindful of the position of the rotating blade in relation to one's legs.
   C. Grasp handle and adjust lever before raising cover plate.
   D. If you prefer, a more direct but slightly more risky method may be applied to the apparatus.

2. Which of the following is LEAST likely to be the subject of a formal report? 2.____
   A. Explorations of adopting new internal procedures within a department
   B. Research into new developments in a field
   C. Explorations of the advisability of launching a new product or an expanded service
   D. An end-of-year review of developments within an organization

3. Which of the following are most clearly eligible for copyright protection? 3.____
   A. Listings of contents
   B. Slogans
   C. Glossaries
   D. Titles

4. When writing a job description, each of the following is a useful rule or guideline EXCEPT 4.____
   A. focusing on the tasks involved rather than the people who do the work
   B. listing duties in increasing order of importance
   C. beginning each duty statement with a verb
   D. before starting, reviewing existing job descriptions

5. Which of the following editing terms is associated with content rather than language? 5.____
   A. Consistency
   B. Clarity
   C. Correctness
   D. Accuracy

6. If a report includes a list of abbreviations and symbols, this list should typically appear in the 6.____
   A. front matter, directly after the preface
   B. front matter, directly after the table of contents
   C. back matter, directly before the appendix or appendices
   D. back matter, directly before the index

7. Every feasibility report should end with a 7.____
   A. list of alternatives B. recommendation
   C. conclusion D. description of methodology

8. Which of the following symbols is referred to as a solidus? 8.____
   A. / B. @ C. Σ D. ¶

9. In formatted tables, the column headings are carried in the element known as the 9.____
   A. horizon B. stub C. rule D. boxhead

10. Of the following methods for outlining a technical document, which is usually the most time-consuming and least effective? 10.____
    A. Random list B. Idea wheel
    C. Formal academic outline D. Computer-screen outline

11. For which of the following types of documents would an index probably be most useful? 11.____
    A. Investigative reports B. Progress reports
    C. Government proposals D. Manuals

12. Which of the following types of illustrations is generally most appropriate for relating parts to a whole? 12.____
    A. Multiple line graph B. Column graph
    C. Pictograph D. Pie graph

13. More than other technical documents, laboratory reports emphasize 13.____
    A. the equipment and the procedures used in an investigation
    B. problems encountered in the investigation
    C. the statement of the investigation's purpose and scope
    D. specific recommendations based upon the report's conclusion

14. What is the term for the device which explains shading, colors, line styles, or other visual ways to distinguish the elements on a graph from one another? 14.____
    A. Callout B. Legend
    C. Banner head D. Nut graf

15. Which of the following is a rule of thumb for preparing pie graphs? 15.____
    A. If the wedges are shaded, do so clockwise and from light to dark.
    B. Keep all labels off to one side.
    C. When possible, begin at the 12 o'clock position and sequence the wedges clockwise, from smallest to largest.
    D. A single percentage point should be considered to be about 5° on the circle.

16. As a general rule, when a report contains more than _____ table(s), they should be listed, along with their titles and page numbers, in a separate section entitled *List of Tables*. 16.____
A. 1 B. 3 C. 5 D. 10

17. Bar graphs are commonly used to show each of the following EXCEPT quantities of 17.____
A. different items for the same time period
B. the same item at different times
C. different items at different times
D. the different parts of an item that make up the whole

18. The majority of technical reports, especially long ones, should be organized as 18.____
A. descending presentations
B. chronological descriptions
C. ascending presentations
D. comparisons

19. Which of the following is a general rule or guideline for the use of heads in technical reports? 19.____
A. If a table of contents is to be used, it should not simply list heads and subheads.
B. All heads at one level should be parallel in structure.
C. If one unit at a particular level is subdivided by lower-level head, every other unit at that level must also include lower-level heads.
D. Second-level heads should appear in all capitals.

20. Which of the following sentences contains a gerund? 20.____
A. Planning to edit on the computer, he established templates before beginning his work.
B. The artist designing the tables made notes in the margins of the typescript.
C. While the managers were deciding upon a way to streamline the procedure, the manual was already being written.
D. The team had already decided that editing for content would be the responsibility of an independent contractor.

21. A hypertext system that is modifiable is described as 21.____
A. nodular B. standard C. robust D. modest

22. Each of the following is a guideline for titling a formal report EXCEPT 22.____
A. indicating the date of the report
B. not putting titles in sentence form
C. avoiding the use of abbreviations and acronyms
D. not using the word *report* in the title

23. If a writer is leading up to an unpleasant or unexpected conclusion, which method of organization is she most likely to use? 23.___
   A. Comparison
   B. Decreasing order of importance
   C. Chronological
   D. Increasing order of importance

24. The copymarking symbol shown at the right, appearing in the corrected proofs of a document, is an indication that in revising the author should 24.___

   A. move the number into subscript: E_2
   B. write out the number: E two
   C. move the letter into superscript: E^2
   D. delete the number

25. Which of the following is NOT considered to be an effective means of addressing the needs of secondary readers? 25.___
   A. Arranging ideas in a need-to-know format
   B. Designing different sections for different readers
   C. Specifically labeling certain parts as too challenging or extraneous for certain readers
   D. Preparing multiple documents

KEY (CORRECT ANSWERS)

1. C
2. A
3. C
4. B
5. D

6. A
7. B
8. A
9. D
10. C

11. D
12. D
13. A
14. B
15. A

16. C
17. C
18. A
19. B
20. D

21. C
22. A
23. D
24. C
25. C

EXAMINATION SECTION

TEST 1

DIRECTIONS: Each question or incomplete statement is followed by several suggested answers or completions. Select the one that BEST answers the question or completes the statement. *PRINT THE LETTER OF THE CORRECT ANSWER IN THE SPACE AT THE RIGHT.*

1. You are preparing a press release announcing a cornerstone laying ceremony for a housing project named after a prominent New Yorker. You desire to include in this press release some information about this person's contributions to public housing. 1.____
Of the following sources which are available to you, the BEST one to go to in order to obtain this information is
   A. the index and issues of a local newpaper obtainable in the public library
   B. the World Almanac
   C. a book on the history of public housing
   D. a biography of the individual

2. You have been assigned to prepare a press release announcing the issuance of applications for apartments at a new city housing project. Of the following items of information, the one which it is LEAST important to include in such a press release is the 2.____
   A. average cost per apartment
   B. rental charges per room
   C. number of apartments in the project
   D. special facilities available at the project

3. The chairman of the housing authority has asked you to assist him in preparing a speech he is to deliver at the ground-breaking ceremonies for the Authority's 100th permanent housing project. Members of the city, state and federal administrations will be present, as well as the press and the general public. 3.____
Of the following, the theme which you should emphasize *most* in this speech is the
   A. role of the chairman in expediting housing progress
   B. failure of private industry to provide for the housing needs of low-income families
   C. steady march toward the elimination of the city's slums
   D. benefits of living in a democracy

4. You have been assigned to prepare an information brochure which is to be distributed to public welfare clients. After preliminary study you find that the value of this brochure will be greatly increased if it is prepared in such a way as to include several pictorial illustrations. You do not have the skill necessary to prepare these illustrations. 4.____
Of the following, the *best* action for you to take is to
   A. prepare the brochure in such a manner as not to require illustrations
   B. ask another employee, who can do the illustrations, to assist you by doing the illustrations

C. prepare the brochure with such illustrations as you are able to draw
D. find out first whether someone will be available to do the illustrations for you

5. The deputy commissioner of the Department of Welfare has asked you to assist him in preparing a speech. The deputy commissioner is to represent the department as the guest of honor at a banquet given by a civic organization. 5.____
Of the following, the *most desirable* action for you to take before beginning to write this speech is to
A. consult the civic organization to secure background material
B. arrange for a brief conference with the deputy commissioner in order to determine his wishes as to the general tone and content of the speech
C. secure a copy of a speech delivered on a similar previous occasion and closely model your speech after it
D. find out whether the speech will be broadcast

6. Of the following, it is generally *most desirable* that informational material written for reading by public welfare clients be 6.____
A. brief and concise
B. easily understood
C. couched in correct technical terms
D. easy to translate into foreign languages

7. Suppose that you are assigned to release department information to reporters for the metropolitan press. Of the following, the LEAST desirable practice for you to adopt in this assignment is 7.____
A. as a general rule, release information in written form only
B. set regular dates for the release of department news insofar as possible
C. secure clearance for the issuance of all written releases
D. release information first to reporters for newspapers which give the best coverage to department news

8. A letter from a private citizen, complaining about a department policy which has worked a hardship on him, has been referred to you for reply. The citizen asks that this policy be changed. 8.____
In answering this letter, it would be *best* to give major emphasis to
A. an explanation of the reasons which make such a policy necessary
B. pointing out that the department regulations cannot be revised to suit each individual case
C. stating that the operations of any large organization must result in some hardships
D. inducing the individual to come into the office where the matter can better be dealt with in a face-to-face interview

9. Suppose you are assigned to prepare the annual report for your department. Each bureau has been asked to submit a written report on its activities for the preceding year. Of the following, the *most desirable* action for you to take in carrying out this assignment is to 9.____
   A. return to the bureau heads for revision those reports which, in your opinion, contain unimportant material
   B. rewrite the material submitted by the bureaus to secure improved style without changing content
   C. arrange a conference with the bureau heads to discuss the reports they are to submit
   D. write an introduction and conclusion and let the reports of the bureaus constitute, unaltered, the body of the annual report

10. You have been assigned by your supervisor to do the preliminary editing of material written by other information assistants. After a week in this assignment you evaluate the material submitted by one information assistant as of lower quality than that of the others. Of the following, the *best* action for you to take is to 10.____
   A. analyze his work with the other information assistants
   B. continue to edit his work without comment at this time
   C. suggest to him that he take a refresher course in writing
   D. recommend his transfer to less original work

11. You have completed gathering the necessary data for a routine newspaper release you are to write. The *most desirable* step for you to take next is to 11.____
   A. write a first draft of the release
   B. work out a plan for the release, including the beginning, the main points, and the ending
   C. develop a suitable title and then begin to write
   D. have someone familiar with the field check the accuracy of the data which you have gathered

12. Of the following writing techniques, the one which is generally LEAST effective for making written matter more forceful is the 12.____
   A. repetition of a key word or phrase
   B. liberal use of exclamation points, capitalization, underlining, and other similar devices
   C. use of the verbs in the active voice, rather than the passive voice
   D. use of a brief sentence, rather than a longer one, to express the same idea

13. The use of anecdotes and other verbal illustrations in writing is desirable *primarily* because 13.____
   A. this is a good way of showing the author's interest in his subject
   B. the reader will remember the anecdotes
   C. the illustrations will help the reader to remember the author's main idea
   D. the illustrations will entertain the reader

14. The technique of directly addressing the reader of a novel or short story 14.____
    A. has been gaining favor steadily during recent years
    B. has been the prevailing practice for a long time
    C. is more common in popular fiction than in literary fiction
    D. is considered out-of-date today

15. The one of the following which is considered LEAST important is good newswriting is 15.____
    A. complete accuracy of names and addresses
    B. full identification of sources of information
    C. strict chronological order of presentation
    D. avoiding the use of editorial statements

16. Of the following, the *best* procedure to follow when writing an article to be read by experts is to 16.____
    A. avoid the technical terms as far as possible
    B. explain the technical terms the first time they are used
    C. use the technical terms of the experts
    D. use your literary judgment as to whether to use the technical terms

17. Of the following, the purpose for which it is LEAST important for a writer to have a large vocabulary is to 17.____
    A. give him a wider choice of synonyms and antonyms
    B. enable him to express himself in a sophisticated language
    C. improve his reading comprehension
    D. make his writing more exact

18. "The family lived in a small edifice on Maple Street." The preceding sentence involves a 18.____
    A. good choice of words
    B. poor choice of words because an "edifice" is large rather than small
    C. poor choice of words because the word "edifice" is obsolete
    D. poor choice of words because the word "edifice" is unfamiliar to the average reader

19. In fiction, the *best* way of acquainting the reader with the traits of the characters is through 19.____
    A. action
    B. dialogue and description
    C. action and dialogue
    D. dialogue

20. Subheads in an informal pamphlet 20.____
    A. are a matter of individual preference
    B. are appropriate *only* if the subject readily breaks itself down into separate sections
    C. should be used because the pamphlet will be easier to read
    D. should NOT be used because they look "textbookish"

21. The length of an average paragraph should 21.____
   A. be about 300 words
   B. harmonize with other elements of a writer's style
   C. not fall below 60 words
   D. vary according to each writing assignment

22. In writing for today's readers, the one of the following which would be LEAST suitable as a literary model for imitation is 22.____
   A. Abraham Lincoln
   B. Samuel Johnson
   C. Mark Twain
   C. Benjamin Franklin

23. Fictitious characters in factual writing should 23.____
   A. be disguised to make them appear real
   B. be given names rather than symbols
   C. be given symbols, such as A, B, and C, rather than names
   D. not be used

24. "Cliches should be avoided in writing." The one of the following which is NOT a cliche is 24.____
   A. "every Tom, Dick, and Harry"
   B. "left no stone unturned"
   C. "outrageous possibilities"
   D. "strike while the iron is hot"

25. Recent polls of the general public indicate that from 20% to 80% of the American people are unacquainted with such items of general information as the United Nations and the Marshall plan. Of the following, the *most probable* cause for this lack of knowledge is that 25.____
   A. people generally don't read enough to grasp this information
   B. most people don't know anything about current events or international relations
   C. the schools avoid the teaching of controversial subjects
   D. this news was not dealt with in the newspapers read by the people polled

26. The "Readers' Guide to Periodical Literature" is 26.____
   A. a digest of magazine articles
   B. a literary magazine
   C. an index of magazine articles
   D. an annual guide to magazine

27. Of the following metropolitan newspapers, the one which has the largest daily circulation is the 27.____
   A. Daily News
   B. Newsday
   C. New York Times
   D. New York Post

28. The cost-of-living index is computed by the 28.____
   A. Bureau of Internal Revenue
   B. Bureau of Labor Statistics
   C. Federal Security Agency
   D. National Bureau of Standards

29. The trend revealed by the U.S. census with regard to the population of metropolitan areas may *best* be described as one of 29.____
   A. little change
   B. marked decline
   C. sharp growth
   D. shift from the center to the suburbs

30. In the United States, agreements that prohibit or restrict the sale of real estate to particular racial groups 30.____
   A. are a very common legal practice
   B. are commonly practiced in the South
   C. were declared legal and enforceable by the U.S. Supreme Court
   D. were declared legally unenforceable by the U.S. Supreme Court

31. Of the following sentences, the one which is poorly written because it contains a "dangling construction" is 31.____
   A. After waiting half an hour for the bus, I remembered that I had no money for carfare.
   B. Having returned from our vacations, the supervisor made reassignments.
   C. Smiling pleasantly, she acknowledged the applause of the audience.
   D. Walking over to him, I introduced myself and offered to help him catch his assailant.

QUESTIONS 32-36.
Questions 32-36 consist of three sentences each. For each question select the sentence which contains NO error in grammar or usage and write the capital letter preceding that sentence in the correspondingl numbered space on your answer sheet.

32. A. Be sure that everybody brings his notes to the conference. 32.____
   B. He looked like he meant to hit the boy.
   C. Mr. Jones is one of the clients who was chosen to represent the district
   D. All are incorrect

33. A. He is taller than I. 33.____
   B. I'll have nothing to do with these kind of people.
   C. The reason why he will not buy the house is because it is too expensive.
   D. All are incorrect

34. A. Aren't I eligible for this apartment. 34.____
   B. Have you seen him anywheres?
   C. He should of come earlier.
   D. All are incorrect

35. A. He graduated college in 1982. 35.____
   B. He hadn't but one more line to write.
   C. Who do you think is the author of this report?
   D. All are incorrect

36. A. I talked to one official, whom I knew was fully impartial. 36.____
B. Everyone signed the petition but him.
C. He proved not only to be a good student but also a good athlete.
D. All are incorrect

QUESTIONS 37-40.
Questions 37-40 consist of three sentences each. For each item, select the sentence which contains NO error in word usage and write the capital letter preceding that sentence in the correspondingly numbered space on your answer sheet.

37. A. Every year a large amount of tenants are admitted to housing projects. 37.____
B. Henry Ford owned around a billion dollars in industrial equipment.
C. He was aggravated by the child's bead behavior.
D. All are incorrect

38. A. Before he was committed to the asylum he suffered from the illusion that he was Napoleon. 38.____
B. Besides stocks, there were also bonds in the safe.
C. We bet the other team easily.
D. All are incorrect

39. A. Bring this report to your supervisor immediately. 39.____
B. He set the chair down near the table.
C. The capitol of New York is Albany.
D. All are incorrect

40. A. He was chosen to arbitrate the dispute because everyone knew he would be disinterested. 40.____
B. It is advisable to obtain the best council before making an important decision.
C. Less college students are interested in teaching than ever before.
D. All are incorrect

KEY (CORRECT ANSWERS)

1. D	11. B	21. D	31. B
2. A	12. B	22. B	32. A
3. C	13. C	23. B	33. A
4. D	14. D	24. C	34. D
5. B	15. C	25. A	35. C
6. B	16. C	26. C	36. B
7. D	17. B	27. A	37. D
8. A	18. B	28. B	38. B
9. C	19. C	29. D	39. B
10. B	20. C	30. D	40. A

TEST 2

DIRECTIONS: Each question or incomplete statement is followed by several suggested answers or completions. Select the one that BEST answers the question or completes the statement. *PRINT THE LETTER OF THE CORRECT ANSWER IN THE SPACE AT THE RIGHT.*

1. "Study your audience and slant your writing toward it." Of the following, the *best* procedure to adopt in applying this principle is to 1.____
   A. estimate the intelligence of your audience and write accordingly
   B. use the simplest possible prose style
   C. write about the things you believe your audience wants to read, rather than the things you would prefer to write about
   D. write about what you want to say in the form that is most likely to appeal to your audience

2. "The first rule for giving your writing 'punch' is to take the most important idea and save it until the end of the sentence." 2.____
   Of the following sentences, the one which *best* illustrates this principle is:
   A. After they had notified the police, and had searched the entire neighborhood for hours, they found the little girl in the attic, sleeping peacefully.
   B. The enemy has destroyed the lives of our people, plundered our seas, ravaged our coasts, and burnt our towns.
   C. The thief had stolen the top secret report, broken open the safe, and rifled the desk.
   D. The tornado left ruin and death in its wake and tore down every building in the village.

3. "America has been built by the cooperative effort of many different kinds of people, working together." 3.____
   In the preceding sentence, a word or phrase which is NOT made superfluous by the use of another word or phrase of similar meaning is
   A. different B. kinds of
   C. many D. working together

4. "The company did so well in 1980 that, at the end of the year, it gave each employee a carton of cigarettes, a bottle of wine, and -- a $100 bond." 4.____
   In the preceding sentence, the dash
   A. adds more force to the words which follow
   B. detracts from the force of the words which follow
   C. is an illustration of the improper use of punctuation
   D. neither adds nor detracts from the force of the words which follow

5. A letter written by another information assistant begins with this sentence: "We beg to acknowledge yours of the 23d inst." It then goes on to reply directly to the matters raised in the letter of the 23d. 5.____
If you are assigned to edit this letter, the most desirable action of the following for you to take is to
   A. change the first sentence to read: "We beg to acknowledge yours of the 23d inst. and in reply wish to state that..."
   B. leave the first sentence as it is
   C. leave the first sentence unchanged but add another immediately following summarizing what the letter of the 23d inquired about
   D. omit the first sentence in its entirety

6. "Write as you talk" is an axiom now widely accepted by newspapermen. Newspaper readers have a better chance of grasping the news if it is told to them simply and clearly. 6.____
The *most direct* implication of the preceding statement is that
   A. an axiom is a statement whose truth is generally accepted by everyone
   B. flowery or complicated language should generally be avoided in newspaper reporting
   C. newspaper readers are no different from newspaper reporters
   D. the use of ungrammatical constructions is sometimes justified in writing for the newspapers

7. "Nowadays, lack of information usually goes hand in hand with little education; similarly, lack of information also usually goes hand in hand with low income. So, if you are writing for people in the lower income brackets or people who haven't gone to college, it's a good guess that they won't have much background knowledge." 7.____
The preceding statement implies *most directly* that
   A. little education has always been negatively correlated with little information
   B. poor people are usually not well-informed
   C. people who have not gone to college are in the lower income brackets
   D. writing for the poor and uneducated is more difficult than writing for the rich and well-educated

8. "Prices of building materials are, in the aggregate, more rigid than those of other commodities. Concentration of control over the supply of goods is frequently advanced as the explanation for price rigidities in general and for building materials in particular." 8.____
According to the preceding statement,
   A. increased demand and concurrent fixed supply are frequently responsible for increased prices of building materials
   B. in the aggregate, the high cost of building materials contributes substantially to the high cost of new housing construction
   C. the cost of most articles is generally more flexible than the cost of articles required in the construction of new buildings
   D. the existence of faulty methods of distribution is often advanced as an argument to explain price inequities

9. "In undertaking a new development, the builder first decides upon the price or rental range of the dwellings he proposes to construct. Then, after roughly estimating the cost of the selected structure, he tries to find land at suitable prices." 9.____
According to the preceding statement,
   A. after a new development is completed, the builder adds up his construction and land costs and fixes the price of the individual house accordingly
   B. it is difficult to predict the probable cost of a new dwelling unit because of constant fluctuation in the cost of building materials
   C. land costs influence the selling price of dwellings least
   D. the selling price of a house is usually determined before construction is begun

10. "A construction program initiated by public agencies better protects the home buyer and insures the greater soundness of the neighborhood." 10.____
According to the preceding statement,
   A. a home buyer is more confident of the safety of his investment if he is given to understand that the neighborhood will not change
   B. a public agency is more responsible in construction programs than a private builder could hope to be
   C. since a public agency can, if necessary, control the development of a neighborhood through zoning laws, public housing is more desirable
   D. to insure the soundness of a neighborhood it is more effective to have the building of new homes planned by public agencies

11. "To achieve sound planning we cannot rely on educating the builder to the fact that what is good for the public will be ultimately good for him, for his interest is usually short term and the pattern in which he functions is not set up for voluntary reform. 11.____
According to the preceding statement,
   A. a builder is not interest in educating the public to its ultimate benefits
   B. builders whose interests are usually of short duration can be educated to set up voluntary reforms
   C. since a builder's interest in any property is usually of short duration, he will voluntarily function for public benefit
   D. we cannot rely on educating a builder to the fact that public benefit is to his advantage in the long run

12. "If cities had a long range objective, if they had plans showing the expected line of growth, plans for their future schools and parks, their houses and their locations, their industries and their locations, their future transportation facilities and their utilities, then with the advent of an emergency requiring government spending they could channel the expenditures and step up the program along the lines of the larger long term plans." 12.____
According to the preceding statement,
   A. a city wishing to eliminate slums can with proper planning

take advantage of an emergency requiring the channeling of expenditures
B. an emergency requires the channeling of expenditures so that greater efficiency can be shown in planning
C. cities which have long range plans can make better use of the funds spent by the government during a depression
D. long range objectives help a city to devise new plans for the development of parks, schools and other public improvements at a considerable saving

13. "Increment or decrement in city income hangs largely upon the maintenance of the values and valuations of real property, upon the quantity of new improvements that go into the city, upon the profitableness of real estate, upon the advent of booms and depressions, and upon the flow of people into or out of the city." 13.____
According to the preceding statement,
A. a boom or a depression has a marked effect on the flow of people into or out of a city
B. new improvements that go into a city enhance the profitableness of real estate
C. real estate values, which form the major basis of a city's taxation, are the sources of city salaries
D. the variation of a city's income depends on the values of the real estate in the city

14. "The institution of the family is a vitally important part of all human societies, but in modern society, particularly, various organized services have developed that enable some people to secure some of the most essential benefits of family life without belonging to a family group." 14.____
Of the following, the LEAST valid inference on the basis of the preceding statement is that
A. people who are not part of a family unit can obtain most of the essential benefits of family life by contacting an appropriate social agency
B. present day society offers an opportunity to some who are not members of a family unit to share in some of the benefits of family living
C. the institution of the family is not native to modern society alone
D. to obtain the benefits of family life it is usually necessary to belong to a family group

15. "Reform organizations seek, as a rule, to bring about a specific economic or political change; social work agencies are usually occupied with the task of meeting existing situations in the lives of particular individuals or groups." 15.____
According to the preceding statement,
A. a reform organization is concerned with helping the individual by changing some factor in the environment which the individual feels is too arduous to accept
B. a reform organization is not concerned with the ability of the individual to meet his social responsibilities
C. social work agencies are not concerned with any specific economic or political change because this does not involve the individual's personal adjustment
D. social workers are primarily concerned with helping their clients to meet current living conditions

16. "Adequate facilities for education, recreation and health must be provided for children, and social conditions created that promote the child's development into a law-abiding citizen. It is not the task of social work to provide these facilities but to direct children to them and to help them to use these facilities." 16.____
Of the following, the *most accurate* statement on the basis of the preceding statement is that
   A. a child who does not have adequate educational, recreational and health facilities will develop into a poor citizen
   B. the education of the public to the importance of providing adequate facilities for children is primarily the social worker's responsibility
   C. the proper use of leisure time by children is an important aspect of the social worker's job
   D. the three most important needs of a child which must be satisfied first are those of education, recreation and health

17. "Social workers start from the assumption that preservation of the family as the basic unit of social living is their accepted objective. In view of the frequency of divorce and the breakdown of authority in the home, social work now makes articulate its concern for family integrity." 17.____
According to the preceding statement,
   A. failure to keep the family as a basic unit leads to a breakdown of authority in the home, upsetting family integrity
   B. in extreme cases where divorce is inevitable a social worker must accept the breakdown of the family unit
   C. social workers are primarily concerned with keeping a family together as a basic entity of social living
   D. the importance of the family to society has been demonstrated by experience with children who have been institutionalized

18. "The marked change in the spirit in which social work is carried on is evidenced in the adoption of business methods of organization, including centralized purchasing of supplies for social agencies, cost accounting, careful budgeting and auditing of accounts, evaluation of methods and publication of reports. Trained personnel for defined jobs is increasingly sought, and there is appreciation of the differentiated abilities required in the social agency." 18.____
According to the preceding statement,
   A. it is apparent that the adoption of business methods of organization has resulted in a chage in the method of preparing case work reports
   B. social work agencies that train people for definite jobs achieve savings in social work that approximate those of business organization.
   C. social work now uses current business procedures in carrying forward the pruposes of a social agency
   D. trained personnel in social work are responsible for the adoption of business methods of procedure

19. "Basic to the functioning of the professional social worker is an understanding of human personality and of the world we live in." 19.____

The one of the following which is the *most accurate* statement on the basis of the preceding quotation is that

A. a social worker must be familiar with human behavior in order to be able to perform his work properly
B. a social worker who understands human personality is able to function better as a citizen of the world
C. social work may be classified as a profession because, for its proper performance, a basic understanding of the social and biological sciences is required
D. through his daily contact with his clients a social worker will obtain a better understanding of the world he lives in

QUESTIONS 20-24.

Questions 20-24 each consist of three words. For each item, select the word which is INCORRECTLY spelled and write the capital letter preceding that work in the correspondingly numbered space on your answer sheet.

20. A. achievment B. maintenance C. questionnaire D. all are correct 20.____

21. A. prevelant B. pronunciation C. separate D. all are correct 21.____

22. A. permissible B. relevant C. seize D. all are correct 22.____

23. A. corroborate B. desparate C. eighth D. all are correct 23.____

24. A. exceed B. feasibility C. psycological D. all are correct 24.____

QUESTIONS 25-29.

Use the material which follows in answering questions 25-29. Copy I on the following page is an accurate copy of material which is to be prepared for the printer. Copy II of this material contains a number of typographical errors. Compare Copy II with Copy I and find the typographical errors. Every group of five lines in Copy II is numbered. Indicate the number of typographical errors in each group of five lines of Copy II by writing in the correspondingly numbered space on the answer sheet the capital letter preceding the best of the following alternatives:

A. no errors
B. 1-2 errors
C. 3-4 errors
D. 5 or more errors

COPY I

Parcel 1. Beginning at a point formed by the intersection of the northerly side of 73rd avenue with the westerly side of Francis Lewis boulevard as said streets are indicated upon the final map of the borough of Queens known as Alteration Map No. 2831 adopted by the board of estimate on May 15, 1941; running thence northerly along the westerly side of Francis Lewis boulevard following a curve having a radius of 8,053 feet for a distance of 585.15 feet; thence northerly along the westerly side of Francis Lewis boulevard in a straight line for a distance of 687.43 feet; thence northerly along the westerly side of Francis Lewis boulevard and its prolongation following a curve having a radius of 5,677 feet for a distance of 509.79 feet to the old southerly side of North Hempstead turnpike as formerly laid out and as shown discontinued upon the aforementioned final city map; thence easterly along said southerly side of North Hempstead turnpike for 110.12 feet to the easterly side of Francis Lewis boulevard; thence southerly along the easterly side of Francis Lewis boulevard following a curve having a radius of 5.783 feet for a distance of 489.20 feet; thence southerly along the easterly side of Francis Lewis boulevard in a straight line for a distance of 687.43 feet; thence southerly along the easterly side of Francis Lewis boulevard following a curve having a radius of 7,947 feet for a distance of 572.90 feet to the northerly side of 73rd avenue.

COPY II

25. Parcel 1: Beginning at point formed by the intersection of the northerly side of 73d Avenue with the westerly side of Francis Lewis boulevard as said streets are indicated upon the final map of the borough of Queens known as Alteration Map No. 2831 adapted by the board of estimate on May 15, 1941; 25.____

26. running thence northerly along the westerly side of Francis Lewis boulevard following a curve having a radius of 8,053 feet for a distance of 585.15 feet; thence northerly along the westerly side of Francis Lewis boulevard in a straight line for a distance of 687.43 feet; thence northerly along 26.____

27. the westerly side of Francis Lewis boulevard and its prolongation following a curve having a radius of 5.677 feet for a distance of 509.79 feet to the old southerly side of North Hempstead Turnpike as formerly laid out and is shown discontinued upon the aforementioned final city map; thence easterly 27.____

28. along said southerly side of North Hempstead turnpike for 1101.2 feet to the easterly side of Francis Lewsis boulevard; thence southerly along the easterly side of Francis Lewis boulevard following a curve having a radius of 5.783 feet for a distance of 489.20 feet; thence southerly along the easterly side 28.____

29. side of Francis Lewis boulevard in a straight line for a distance of 687.43 feet; thence southerly along the easterly side of Francis Lewis boulevard following a curve having a radius of 7,947 feet for a distance of 572.90 feet to the northerly side of 73d avenue. 29.____

30. "He described a hypothetical situation to illustrate his point." In the preceding sentence, the word "hypothetical" means *most nearly* 30.____
   A. actual
   B. theoretical
   C. typical
   D. unusual

31. "I gave tacit approval to my partner's proposed business changes." In the preceding sentence, the word "tacit" means *most nearly* 31.____
   A. enthusiastic
   B. partial
   C. silent
   D. written

32. "Jones was considered an "astute lawyer by the members of his profession." In the preceding sentence, the word "astute" means *most nearly* 32.____
   A. clever
   B. persevering
   C. poorly trained
   D. unethical

33. "There were intimations even in early days of the way in which he would go." In the preceding sentence, the word "intimations" means *most nearly* 33.____
   A. hints
   B. patterns
   C. plans
   D. purposes

34. "His last book was published posthumously." In the preceding sentence, the word "posthumously" means *most nearly* 34.____
   A. after the death of the author
   B. printed free by the publisher
   C. without a dedication
   D. without royalties

35. "When he was challenged, he used every known subterfuge." In the preceding sentence, the word "subterfuge" means *most nearly* 35.____
   A. evasion to justify one's conduct
   B. means of attack to defend one's self
   C. medical device
   D. unconscious thought

36. "His partner suggested a course of action that would alleviate the difficulties which confronted him." In the preceding sentence, the word "alleviate" means *most nearly* 36.____
   A. correct
   B. lessen
   C. remove
   D. solve

37. "Among the applicants for the new apartment white collar workers were preponderant." In the preceding sentence, the word "preponderant" means *most nearly* 37.____
   A. considered not eligible
   B. in evidence
   C. superior in number
   D. the first to apply

38. "The captain gave a lucid explanation of his plans for the coming campaign." In the preceding sentence, the word "lucid" means *most nearly* 38.____
   A. clear
   B. graphic
   C. interesting
   D. thorough

39. "He led a sedentary life." In the preceding sentence, the word "sedentary" means *most nearly* 39.____
   A. aimless
   B. exciting
   C. full
   D. inactive

40. "His plan for the next campaign was very plausible." In the preceding sentence, the word "plausible" means *most nearly* 40.____
   A. appropriate
   B. believable
   C. usable
   D. valuable

KEY (CORRECT ANSWERS)

1. D	11. D	21. A	31. C
2. A	12. C	22. D	32. A
3. C	13. D	23. B	33. A
4. A	14. A	24. C	34. A
5. D	15. D	25. C	35. A
6. B	16. C	26. A	36. B
7. B	17. C	27. C	37. C
8. C	18. C	28. B	38. A
9. D	19. A	29. B	39. D
10. D	20. A	30. B	40. B

EXAMINATION SECTION

TEST 1

DIRECTIONS: Each question or incomplete statement is followed by several suggested answers or completions. Select the one that *BEST* answers the question or completes the statement. *PRINT THE LETTER OF THE CORRECT ANSWER IN THE SPACE AT THE RIGHT.*

1. The informed editorial assistant knows that the difference between a copy reader and a proofreader is that the copy reader 1.____
   A. checks copy against type proofs
   B. checks galleys
   C. edits material submitted by writers
   D. holds copy for the proofreader

2. *Style* to a copy editor means 2.____
   A. following a set pattern when rules of spelling and punctuation are equivocal
   B. following the rules of formal grammar
   C. making sure that the writing is not *elegant*
   D. making sure that the writing is polished

3. Before it is sent to the composing room, each page of copy *MUST* be 3.____
   A. copy edited
   B. galley proofed
   C. given a folio number
   D. proofread

4. When a long paragraph in a manuscript is to be divided into two paragraphs, it should be done 4.____
   A. by circling the first word of the new paragraph
   B. by marking an ¶ around the beginning of the new paragraph
   C. by marking dividing spot with caret and writing note to printer in margin
   D. with paste pot and shears at point of division

5. To a copy editor, the word *more* means 5.____
   A. facts are insufficient
   B. linotypists need copy to keep machines running
   C. story goes to next take
   D. story is not long enough

6. To abbreviate the word *Company*, as in Jones Widget Company, a copy editor should 6.____
   A. circle the word
   B. cross out excess letters and put a period over the *m*
   C. cross out the entire word and write *Co.* above it in the space between the lines
   D. cross out the entire word and write *Co.* in margin, running a line to its position

7. In editing copy, it is often necessary to indicate that numerals are to be spelled out. This is done by 7.___
   A. circling the numeral
   B. crossing out the numeral and spelling it out between the lines
   C. crossing out the numeral and spelling it out in margin with a line drawn to its position
   D. drawing a square around the numeral

8. In make-up, pages are laid out on 8.___
   A. dummies B. folios C. plan sheets D. proof sheets

9. Type size is measured 9.___
   A. ems B. inches C. picas D. points

10. A pica is 10.___
   A. just over 1/6 of an inch
   B. just under 1/6 of an inch
   C. 1/72 of an inch
   D. none of the above

11. Columns are measured in 11.___
   A. ems
   B. fractions of a page
   C. picas
   D. points

12. 36 points is 12.___
   A. about an inch
   B. about half an inch
   C. about two inches
   D. none of the foregoing

13. Copy marked to be set *ff* means to make type 13.___
   A. larger B. smaller C. thicker D. thinner

14. The letters in italic type 14.___
   A. are less cursive than in roman type
   B. are more formal than in roman type
   C. are set by hand
   D. slant bottom left to top right

15. Sans-serif type 15.___
   A. has no additional fonts
   B. has no curlicues
   C. has not hangers or risers
   D. is old-fashioned German type

Questions 16-20

DIRECTIONS: Questions 16 through 20 consist of groups of four words.
Select answer A if only ONE word is spelled correctly in a group.
Select answer B if TWO words are spelled correctly in a group.
Select answer C if THREE words are spelled correctly in a group.
Select answer D if all FOUR words are spelled correctly in a group.

16. counterfeit embarass panicky supercede 16.___

17.	benefited	personnel	questionnaire	unparalelled	17.___
18.	bankruptcy	describable	proceed	vacuum	18.___
19.	handicapped	mispell	offerred	pilgrimmage	19.___
20.	corduroy	interfere	privilege	separator	20.___

Questions 21 - 25

DIRECTIONS: For each question numbered 21 through 25, select the option whose meaning is MOST NEARLY the same as that of the numbered item.

21. CONDONE 21.___
A. complete B. condemn C. cooperate D. pardon

22. EXTENUATE 22.___
A. accuse B. excuse C. lengthen D. narrow

23. MORDANT 23.___
A. caustic B. depressed C. dying D. unwholesome

24. SPATE 24.___
A. broad road
B. excessive quantity
C. fish eggs
D. mineral springs

25. TORTUOUS 25.___
A. devious B. foul C. injurious D. painful

KEY (CORRECT ANSWERS)

1.	C	11.	C
2.	A	12.	B
3.	A	13.	C
4.	B	14.	D
5.	C	15.	B
6.	A	16.	B
7.	A	17.	C
8.	A	18.	D
9.	D	19.	A
10.	B	20.	D

21. D
22. B
23. A
24. B
25. A

TEST 2

DIRECTIONS: Each question or incomplete statement is followed by several suggested answers or completions. Select the one that *BEST* answers the question or completes the statement. *PRINT THE LETTER OF THE CORRECT ANSWER IN THE SPACE AT THE RIGHT.*

1. Of the following, the PRINCIPAL reason why most employee house magazines are sent by mail rather than delivered at the plant is to 1.___
   A. reach the families of employees
   B. reduce personnel needed to distribute magazines
   C. shorten time involved in distributing magazines
   D. take advantage of the low mailing rates

2. Press releases received by a newspaper are usually directed to the 2.___
   A. city editor
   B. managing editor
   C. promotion Manager
   D. publisher

3. The agency of the United States Government that supervises radio and television broadcasting is known by the abbreviation 3.___
   A. ABC
   B. FCC
   C. FTC
   D. SEC

4. The names of the sports editors of all state daily newspapers would be found in 4.___
   A. ADVERTISING AGE
   B. Ayer's DIRECTORY OF PERIODICALS
   C. EDITOR AND PUBLISHER INTERNATIONAL YEARBOOK
   D. PUBLIC AFFAIRS INFORMATION JOURNAL

5. Of the following, the type of publicity MOST likely to promote morale of the employees of your department would be a(n) 5.___
   A. article concerning the department written for a technical publication
   B. article in the annual report summarizing the activities of the department
   C. local newspaper article on the accomplishments of the employees of the agency
   D. short *oddity* on your superior carried by AP

6. If you wanted one photograph of a street accident to illustrate the need for improving traffic control at the scene of the accident, you should select a picture that shows 6.___
   A. a close-up of the cars and the victim
   B. a policeman questioning witnesses at the accident scenes
   C. the cars and the victim against the whole intersection of the accident scene
   D. the victim being put into an ambulance

7. The one of the following which is NOT a miniature camera is the 7.____
A. Nikon B. Contax C. Leica D. Rolleiflex

8. To mortise a cut is to 8.____
A. cut out a portion for inclusion of type
B. get it level with surrounding slugs of type
C. return it to engravers for corrections
D. send it to the morgue

9. A picture known as *cheesecake* is an example of 9.____
A. a *mug* shot
B. a scenic shot
C. facsimile reproduction
D. *leg* art

10. A *mat* of a picture is a 10.____
A. copy negative
B. papier mache impression from which plate may be cast
C. picture page dummy from which copies are made
D. print on dull finish paper

Questions 11 - 20

DIRECTIONS: In questions 11 through 20, print on your answer sheet, next to the corresponding question number, the capital letter immediately preceding the word or phrase which is CLOSEST in meaning to that of the capitalized letter.

11. BIBLIOPHILE 11.____
A. appendix
B. library
C. list of references
D. lover of books

12. SACERDOTAL 12.____
A. penitential B. priestly C. reminiscent D. spirtual

13. FLAGELLATE 13.____
A. communicate by signals
B. pillage
C. play the flute
D. scourge

14. SAGA 14.____
A. epoch B. hero C. inscription D. legend

15. APOCRYPHAL 15.____
A. annotated B. orthodox C. unauthentic D. visionary

16. CAVIL 16.____
A. make captious objection
B. punish severely
C. render just praise
D. warn emphatically

17. SHIBBOLETH 17.____
A. seventh year
B. stone block
C. watchword
D. weapon

18. ARRANT 18.____
A. deceitful B. notorious C. overbearing D. parched

19. RUE 19.____
A. abandon B. despair C. repent D. stain

20. MORDANT 20.____
A. caustic B. depressed C. dying D. unwholesome

Questions 21 - 25

DIRECTIONS: Each question or incomplete statement is followed by several suggested answers or completions. Select the one that *BEST* answers the questions or completes the statement. *PRINT THE LETTER OF THE CORRECT ANSWER IN THE SPACE AT THE RIGHT.*

21. Of the following, the grammatically CORRECT sentence is: 21.____
A. Neither the mayor nor the city clerk are willing to talk.
B. Neither the mayor nor the city clerk is willing to talk.
C. Neither the mayor or the city clerk are willing to talk.
D. Neither the mayor or the city clerk is willing to talk.

22. Of the following, the grammatically CORRECT sentence is: 22.____
A. Being that he is that kind of boy, cooperation cannot be expected.
B. He interviewed people who he thought had something to say.
C. Stop whomever enters the building regardless of rank or office held.
D. Passing through the countryside, the scenery pleased us.

23. Of the following, the grammatically CORRECT sentence is: 23.____
A. The childrens' shoes were in their closet.
B. The children's shoes were in their closet.
C. The childs' shoes were in their closet.
D. The childs' shoes were in his closet.

24. Of the following, the grammatically INCORRECT sentence is: 24.____
A. Dissatisfaction with the theoretical bases and practical workings of the general property tax has given rise to two movements of tax reform.
B. Let the book lie on the table.
C. Since the department is reducing its number of employees is not proof that they are not needed.
D. Who do you think will be selected for the position?

25. Of the following, the grammatically INCORRECT sentence is: 25.____
A. Application of the principles discovered during those experiments have been of great value to mankind.
B. Every one of the editorial assistants proved his worth without exception.
C. State regulation of morals aids in the protection of the family.
D. Working when one is tired does not yield the best results.

KEY (CORRECT ANSWERS)

1. A
2. A
3. B
4. C
5. C

6. C
7. D
8. A
9. D
10. B

11. D
12. B
13. D
14. D
15. C

16. A
17. C
18. B
19. C
20. A

21. B
22. B
23. B
24. C
25. A

EXAMINATION SECTION

TEST 1

DIRECTIONS: Each question or incomplete statement is followed by several suggested answers or completions. Select the one that BEST answers the question or completes the statement. *PRINT THE LETTER OF THE CORRECT ANSWER IN THE SPACE AT THE RIGHT.*

1. Of the following, the order in which a piece of local copy is MOST likely to flow is 1.___
   A. city editor, copy reader, compositor
   B. compositor, city editor, copy reader
   C. compositor, copy reader, city editor
   D. copy reader, compositor, city editor

2. The one who edits copy and writes headlines at a newspaper copy desk is USUALLY called a 2.___
   A. copy cutter B. copy holder
   C. rim man D. slot man

3. Of the following, a comprehensive single-volume reference book for general information is 3.___
   A. FACTS ON FILE
   B. the CONGRESSIONAL DIRECTORY
   C. the STATISTICAL ABSTRACT OF THE UNITED STATES GOVERNMENT
   D. the WORLD ALMANAC

4. As a defense against libel, one could claim that a statement was quoted from the CONGRESSIONAL RECORD and, therefore, was a(n) 4.___
   A. indirect quotation B. personal comment
   C. privileged statement D. reporter's prerogative

5. In order to have an illustration continue off the edge of a page, you would instruct the printer to 5.___
   A. bleed B. cutoff C. justify D. mortise

6. The description which accompanies a photograph or diagram is called a(n) 6.___
   A. caption B. flag
   C. overhead D. underhead

7. A headline stretching across all columns of a page is called a 7.___
   A. bank B. banner
   C. cross line D. drop line

8. You would expect a headline that is flush right and left to 8.___
   A. comprise two or more decks
   B. fill the entire line

C. have black em borders at both ends
D. have two ems of white space on both left and right hand sides

9. A headline that carries a story continued from another page is known as a 9.____
A. break B. filler C. jump D. read-in

10. The secondary part of a headline is a 10.____
A. byline B. deck C. slug D. subhead

11. A single piece of type that includes two or more letters is called 11.____
A. dingbat B. linotype
C. logotype D. nonpareil

12. The one of the following terms which indicates that type is jumbled and no longer of use is 12.____
A. fudge B. mash C. montage D. pi

13. Type left over and unused after a magazine has been sent to press is called 13.____
A. folo copy B. make-ready
C. overset D. quoin

14. An upper and lower case crossline reads as follows: Dr. Doe Appointed. The unit count of this crossline is 14.____
A. 14½ B. 16 C. 18 D. 19

15. An incomplete story consists of the following parts: First add, first add insert, insert after first add, lead, precede, second add. 15.____
An additional piece of copy marked *second add insert* is added to complete the story.
In the final sequence of the completed story, this piece of copy would appear as the _____ item.
A. second B. fourth C. fifth D. sixth

16. When copy for a standard-size newspaper is returned with instructions to *cut to 1 col.*, the number of words should be reduced to APPROXIMATELY 16.____
A. 500 B. 1000 C. 1500 D. 2000

17. In copyreading, two straight lines placed under a word indicates that it should be set in 17.____
A. boldface B. capitals and lower case
C. italics D. small capitals

18. When a story is continued on a second page, the copyreader marks the bottom of the first page with 18.____
A. ## B. -30- C. insuff. D. more

19. The copyreader's symbol which is used to indicate that a subhead should be centered is 19. ____

A.  B.  C.  D. 

20. The copyreader's symbol which is used to indicate the start of a paragraph is 20. ____

A.  B.  C.  D. 

Questions 21-25.

DIRECTIONS: Questions 21 through 25 consist of four pairs of words each. Some of the words are spelled correctly; others are spelled incorrectly. For each question, indicate in the space at the right the letter preceding that pair of words in which BOTH words are spelled CORRECTLY.

21. A. hygienic, inviegle B. omniscience, pittance 21. ____
C. plagarize, nullify D. seargent, perilous

22. A. auxilary, existence B. pronounciation, accordance 22. ____
C. ignominy, indegence D. suable, baccalaureate

23. A. discreet, inaudible B. hypocrisy, currupt 23. ____
C. liquidate, maintainance D. transparancy, onerous

24. A. facility, stimulent B. frugel, sanitary 24. ____
C. monetary, prefatory D. punctileous, credentials

25. A. bankruptsy, perceptible B. disuade, resilient 25. ____
C. exhilerate, expectancy D. panegyric, disparate

KEY (CORRECT ANSWERS)

1. A
2. C
3. D
4. C
5. A

6. A
7. B
8. B
9. C
10. B

11. C
12. D
13. C
14. B
15. D

16. B
17. D
18. D
19. C
20. A

21. B
22. D
23. A
24. C
25. D

TEST 2

DIRECTIONS: Each question or incomplete statement is followed by several suggested answers or completions. Select the one that BEST answers the question or completes the statement. *PRINT THE LETTER OF THE CORRECT ANSWER IN THE SPACE AT THE RIGHT.*

1. Spot news stories are USUALLY written 1.____
   A. as feature articles
   B. in inverted pyramid style
   C. to fill inside pages
   D. with chronological organization

2. The Flesch formula is concerned with 2.____
   A. body type size
   B. front page make-up
   C. headline styles
   D. sentence length and number of syllables

3. A good reporter avoids taking extensive notes during an interview because 3.____
   A. accuracy is of secondary importance in reporting an interview
   B. a rewrite man may actually write the story
   C. this may mark the reporter as an amateur
   D. this may distract the person being interviewed

4. A reporter is present at a function where a distinguished person is scheduled to speak. He has a complete advance copy of the text. 4.____
   During the speech, the reporter could BEST use his time to
   A. follow the text to see if the speaker deviates from it
   B. organize the material of the text for later writing
   C. take direct quotations from the text
   D. write headlines for the story

5. A news story written in an inverted pyramid form is one in which the 5.____
   A. climax is reached at the end of the story
   B. climax is reached near the middle of the story
   C. facts are arranged in chronological order of occurrence
   D. facts are arranged in descending order of reader interest

6. In a news story, the first mention of the former Catholic Cardinal of New York should be written as 6.____
   A. Cardinal Terence Cook
   B. Terence Cardinal Cook
   C. The Rt. Rev. Terence Cook
   D. The Very Rev. Terence Cook

7. The second reference to the British statesman in a newspaper story should be written as 7.____
A. Lord Churchill
B. Mr. Churchill
C. Sir Churchill
D. Sir Winston

8. The MOST common type of lead on newspaper stories is the _____ lead. 8.____
A. astonisher
B. quotation
C. summary
D. suspended-interest

9. The one of the following terms which does NOT designate a story accompanying a report of a major news event is 9.____
A. precede
B. shirt tail
C. sidebar
D. subhead

10. A second day story is also known as a 10.____
A. filler
B. flimsy
C. follow copy
D. follow story

11. A phrase or word used on copy to identify additional pages of a news story is called a 11.____
A. headline
B. slot
C. slug
D. stamp

12. A *tear sheet* is a 12.____
A. carbon copy of a story
B. galley proof of a story
C. page proof of a publication
D. printed page from a publication

13. If you were told to *boil* a story, you would 13.____
A. expand with editorial comment
B. present only essential facts
C. try to keep it a *scoop*
D. write a large headline spread

14. If a reporter went to the *morgue*, he would be seeking 14.____
A. carbon copies of an article he had just written
B. clippings on the subject he was writing about
C. galley proofs of a recently completed article
D. incoming press association teletype copy

15. A beat reporter's *future book* is a 15.____
A. chronological listing of expected events on his beat
B. list of news sources on his beat
C. novel he eventually hopes to write
D. schedule of assignments kept for editor of editorial page

16. To a magazine editor, the term *query* means a 16.____
A. letter outlining an article idea
B. personal request for conference with the editor
C. request for an advance
D. rough draft of an article

17. The one of the following who is NOT on the editorial staff of a large metropolitan newspaper is the 17.____
A. copyreader B. photographer
C. proofreader D. rewrite man

18. The conventional name given to a staff member who gathers news on a daily paper and telephones into the office is 18.____
A. bulldog B. cub C. leg man D. slot man

19. Material for future release would probably be marked 19.____
A. F.Y.I. B. H.F.R. C. H.T.K. D. M.E.

20. When a reporter sends a news item by telegraph from out of town at night, he MOST probably would mark it 20.____
A. BOM B. DPR C. EOS D. NPR

21. Of the following newspapers, the one with the LARGEST circulation is the 21.____
A. CHICAGO TRIBUNE B. NEW YORK DAILY NEWS
C. NEW YORK TIMES D. PHILADELPHIA BULLETIN

22. Copyrighted material in the United States can be protected for a MAXIMUM of _____ years. 22.____
A. 16 B. 42 C. 56 D. 99

23. Of the following, the BEST source of biographical information on a famous living American sociologist is 23.____
A. the COLUMBIA ENCYCLOPEDIA
B. the DICTIONARY OF AMERICAN BIOGRAPHY
C. the NATIONAL BIOGRAPHICAL INDEX
D. WHO'S WHO IN AMERICA

24. A book's index is USUALLY prepared from 24.____
A. dummy sheets B. edited copy
C. galley proof D. page proof

25. If you desire to include copyrighted material amounting to well over 2,000 words in an article, you should obtain written permission from the 25.____
A. author
B. copyright holder
C. Library of Congress copyright office
D. publisher

KEY (CORRECT ANSWERS)

1. B	6. B	11. C	16. A	21. B
2. D	7. D	12. D	17. C	22. C
3. D	8. C	13. B	18. C	23. D
4. A	9. D	14. B	19. B	24. D
5. D	10. D	15. A	20. D	25. B

EXAMINATION SECTION

TEST 1

DIRECTIONS: Each question or incomplete statement is followed by several suggested answers or completions. Select the one that BEST answers the question or completes the statement. *PRINT THE LETTER OF THE CORRECT ANSWER IN THE SPACE AT THE RIGHT.*

1. In public agencies, communications should be based PRIMARILY on a 1.____
   A. two-way flow from the top down and from the bottom up, most of which should be given in writing to avoid ambiguity
   B. multidirection flow among all levels and with outside persons
   C. rapid, internal one-way flow from the top down
   D. two-way flow of information, most of which should be given orally for purposes of clarity

2. In some organizations, changes in policy or procedures are often communicated by word of mouth from supervisors to employees with no prior discussion or exchange of viewpoints with employees. 2.____
   This procedure often produces employee dissatisfaction CHIEFLY because
   A. information is mostly unusable since a considerable amount of time is required to transmit information
   B. lower-level supervisors tend to be excessively concerned with minor details
   C. management has failed to seek employees' advice before making changes
   D. valuable staff time is lost between decision-making and the implementation of decisions

3. For good letter writing, you should try to visualize the person to whom you are writing, especially if you know him. 3.____
   Of the following rules, it is LEAST helpful in such visualization to think of
   A. the person's likes and dislikes, his concerns, and his needs
   B. what you would be likely to say if speaking in person
   C. what you would expect to be asked if speaking in person
   D. your official position in order to be certain that your words are proper

4. One approach to good informal letter writing is to make letters sound conversational. 4.____
   All of the following practices will usually help to do this EXCEPT:
   A. If possible, use a style which is similar to the style used when speaking

B. Substitute phrases for single words (e.g., *at the present time* for *now*
C. Use contractions of words (e.g., *you're* for *you are*)
D. Use ordinary vocabulary when possible

5. All of the following rules will aid in producing clarity in report-writing EXCEPT: 5.
   A. Give specific details or examples, if possible
   B. Keep related words close together in each sentence
   C. Present information in sequential order
   D. Put several thoughts or ideas in each paragraph

6. The one of the following statements about public relations which is MOST accurate is that 6.
   A. in the long run, appearance gains better results than performance
   B. objectivity is decreased if outside public relations consultants are employed
   C. public relations is the responsibility of every employee
   D. public relations should be based on a formal publicity program

7. The form of communication which is usually considered to be MOST personally directed to the intended recipient is the 7.
   A. brochure B. film C. letter D. radio

8. In general, a document that presents an organization's views or opinions on a particular topic is MOST accurately known as a 8.
   A. tear sheet
   B. position paper
   C. flyer
   D. journal

9. Assume that you have been asked to speak before an organization of persons who oppose a newly announced program in which you are involved. You feel tense about talking to this group. 9.
   Which of the following rules generally would be MOST useful in gaining rapport when speaking before the audience?
   A. Impress them with your experience
   B. Stress all areas of disagreement
   C. Talk to the group as to one person
   D. Use formal grammar and language

10. An organization must have an effective public relations program since, at its best, public relations is a bridge to change. 10.
    All of the following statements about communication and human behavior have validity EXCEPT:
    A. People are more likely to talk about controversial matters with like-minded people than with those holding other views

B. The earlier an experience, the more powerful its effect since it influences how later experiences will be interpreted
C. In periods of social tension, official sources gain increased believability
D. Those who are already interested in a topic are the ones who are most open to receive new communications about it

11. An employee should be encouraged to talk easily and frankly when he is dealing with his supervisor. 11.___
In order to encourage such free communication, it would be MOST appropriate for a supervisor to behave in a(n)
A. sincere manner; assure the employee that you will deal with him honestly and openly
B. official manner; you are a supervisor and must always act formally with subordinates
C. investigative manner; you must probe and question to get to a basis of trust
D. unemotional manner; the employee's emotions and background should play no part in your dealings with him

12. Research findings show that an increase in free communication within an agency GENERALLY results in which one of the following? 12.___
A. Improved morale and productivity
B. Increased promotional opportunities
C. An increase in authority
D. A spirit of honesty

13. Assume that you are a supervisor and your superiors have given you a new-type procedure to be followed. 13.___
Before passing this information on to your subordinates, the one of the following actions that you should take FIRST is to
A. ask your superiors to send out a memorandum to the entire staff
B. clarify the procedure in your own mind
C. set up a training course to provide instruction on the new procedure
D. write a memorandum to your subordinates

14. Communication is necessary for an organization to be effective. 14.___
The one of the following which is LEAST important for most communication systems is that
A. messages are sent quickly and directly to the person who needs them to operate
B. information should be conveyed understandably and accurately
C. the method used to transmit information should be kept secret so that security can be maintained
D. senders of messages must know how their messages are received and acted upon

15. Which one of the following is the CHIEF advantage of listening willingly to subordinates and encouraging them to talk freely and honestly? 15
It
   A. reveals to supervisors the degree to which ideas that are passed down are accepted by subordinates
   B. reduces the participation of subordinates in the operation of the department
   C. encourages subordinates to try for promotion
   D. enables supervisors to learn more readily what the *grapevine* is saying

16. A supervisor may be informed through either oral or written reports. 16
Which one of the following is an ADVANTAGE of using oral reports?
   A. There is no need for a formal record of the report.
   B. An exact duplicate of the report is not easily transmitted to others.
   C. A good oral report requires little time for preparation.
   D. An oral report involves two-way communication between a subordinate and his supervisor.

17. Of the following, the MOST important reason why supervisors should communicate effectively with the public is to 17.
   A. improve the public's understanding of information that is important for them to know
   B. establish a friendly relationship
   C. obtain information about the kinds of people who come to the agency
   D. convince the public that services are adequate

18. Supervisors should generally NOT use phrases like *too hard*, *too easy*, and *a lot* PRINCIPALLY because such phrases 18.
   A. may be offensive to some minority groups
   B. are too informal
   C. mean different things to different people
   D. are difficult to remember

19. The ability to communicate clearly and concisely is an important element in effective leadership. 19.
Which of the following statements about oral and written communication is GENERALLY true?
   A. Oral communication is more time-consuming.
   B. Written communication is more likely to be misinterpreted.
   C. Oral communication is useful only in emergencies.
   D. Written communication is useful mainly when giving information to fewer than twenty people.

20. Rumors can often have harmful and disruptive effects on an organization. 20.___
Which one of the following is the BEST way to prevent rumors from becoming a problem?
   A. Refuse to act on rumors, thereby making them less believable.
   B. Increase the amount of information passed along by the *grapevine*.
   C. Distribute as much factual information as possible.
   D. Provide training in report writing.

21. Suppose that a subordinate asks you about a rumor he has heard. The rumor deals with a subject which your superiors consider *confidential*. 21.___
Which of the following BEST describes how you should answer the subordinate?
Tell
   A. the subordinate that you don't make the rules and that he should speak to higher ranking officials
   B. the subordinate that you will ask your superior for information
   C. him only that you cannot comment on the matter
   D. him the rumor is not true

22. Supervisors often find it difficult to *get their message across* when instructing newly appointed employees in their various duties. 22.___
The MAIN reason for this is generally that the
   A. duties of the employees have increased
   B. supervisor is often so expert in his area that he fails to see it from the learner's point of view
   C. supervisor adapts his instruction to the slowest learner in the group
   D. new employees are younger, less concerned with job security and more interested in fringe benefits

23. Assume that you are discussing a job problem with an employee under your supervision. During the discussion, you see that the man's eyes are turning away from you and that he is not paying attention. 23.___
In order to get the man's attention, you should FIRST
   A. ask him to look you in the eye
   B. talk to him about sports
   C. tell him he is being very rude
   D. change your tone of voice

24. As a supervisor, you may find it necessary to conduct meetings with your subordinates. 24.___
Of the following, which would be MOST helpful in assuring that a meeting accomplishes the purpose for which it was called?
   A. Give notice of the conclusions you would like to reach at the start of the meeting.
   B. Delay the start of the meeting until everyone is present.

C. Write down points to be discussed in proper sequence.
D. Make sure everyone is clear on whatever conclusions have been reached and on what must be done after the meeting.

25. Every supervisor will occasionally be called upon to deliver a reprimand to a subordinate. If done properly, this can greatly help an employee improve his performance. Which one of the following is NOT a good practice to follow when giving a reprimand? 25.
A. Maintain your composure and temper.
B. Reprimand a subordinate in the presence of other employees so they can learn the same lesson.
C. Try to understand why the employee was not able to perform satisfactorily.
D. Let your knowledge of the man involved determine the exact nature of the reprimand.

KEY (CORRECT ANSWERS)

1. C
2. B
3. D
4. B
5. D

6. C
7. C
8. B
9. C
10. C

11. A
12. A
13. B
14. C
15. A

16. D
17. A
18. C
19. B
20. C

21. B
22. B
23. D
24. D
25. B

TEST 2

DIRECTIONS: Each question or incomplete statement is followed by several suggested answers or completions. Select the one that BEST answers the question or completes the statement. *PRINT THE LETTER OF THE CORRECT ANSWER IN THE SPACE AT THE RIGHT.*

1. Usually one thinks of communication as a single step, essentially that of transmitting an idea. 1.___
Actually, however, this is only part of a total process, the FIRST step of which should be
   A. the prompt dissemination of the idea to those who may be affected by it
   B. motivating those affected to take the required action
   C. clarifying the idea in one's own mind
   D. deciding to whom the idea is to be communicated

2. Research studies on patterns of informal communication have concluded that most individuals in a group tend to be passive recipients of news, while a few make it their business to spread it around in an organization. 2.___
With this conclusion in mind, it would be MOST correct for the supervisor to attempt to identify these few individuals and
   A. give them the complete facts on important matters in advance of others
   B. inform the other subordinates of the identify of these few individuals so that their influence may be minimized
   C. keep them straight on the facts on important matters
   D. warn them to cease passing along any information to others

3. The one of the following which is the PRINCIPAL advantage of making an oral report is that it 3.___
   A. affords an immediate opportunity for two-way communication between the subordinate and superior
   B. is an easy method for the superior to use in transmitting information to others of equal rank
   C. saves the time of all concerned
   D. permits more precise pinpointing of praise or blame by means of follow-up questions by the superior

4. An agency may sometimes undertake a public relations program of a defensive nature. 4.___
With reference to the use of defensive public relations, it would be MOST correct to state that it
   A. is bound to be ineffective since defensive statements, even though supported by factual data, can never hope to even partly overcome the effects of prior unfavorable attacks

B. proves that the agency has failed to establish good relationships with newspapers, radio stations, or other means of publicity
C. shows that the upper echelons of the agency have failed to develop sound public relations procedures and techniques
D. is sometimes required to aid morale by protecting the agency from unjustified criticism and misunderstanding of policies or procedures

5. Of the following factors which contribute to possible undesirable public attitudes towards an agency, the one which is MOST susceptible to being changed by the efforts of the individual employee in an organization is that 5.
A. enforcement of unpopular regulations has offended many individuals
B. the organization itself has an unsatisfactory reputation
C. the public is not interested in agency matters
D. there are many errors in judgment committed by individual subordinates

6. It is not enough for an agency's services to be of a high quality; attention must also be given to the acceptability of these services to the general public. 6.
This statement is GENERALLY
A. *false*; a superior quality of service automatically wins public support
B. *true*; the agency cannot generally progress beyond the understanding and support of the public
C. *false*; the acceptance by the public of agency services determines their quality
D. *true*; the agency is generally unable to engage in any effective enforcement activity without public support

7. Sustained agency participation in a program sponsored by a community organization is MOST justified when 7.
A. the achievement of agency objectives in some area depends partly on the activity of this organization
B. the community organization is attempting to widen the base of participation in all community affairs
C. the agency is uncertain as to what the community wants
D. there is an obvious lack of good leadership in a newly formed community organization

8. Of the following, the LEAST likely way in which a records system may serve a supervisor is in 8.
A. developing a sympathetic and cooperative public attitude toward the agency
B. improving the quality of supervision by permitting a check on the accomplishment of subordinates
C. permit a precise prediction of the exact incidences in specific categories for the following year
D. helping to take the guesswork out of the distribution of the agency

9. Assuming that the *grapevine* in any organization is virtually indestructible, the one of the following which it is MOST important for management to understand is: 9.___
   A. What is being spread by means of the *grapevine* and the reason for spreading it
   B. What is being spread by means of the *grapevine* and how it is being spread
   C. Who is involved in spreading the information that is on the *grapevine*
   D. Why those who are involved in spreading the information are doing so

10. When the supervisor writes a report concerning an investigation to which he has been assigned, it should be LEAST intended to provide 10.___
   A. a permanent official record of relevant information gathered
   B. a summary of case findings limited to facts which tend to indicate the guilt of a suspect
   C. a statement of the facts on which higher authorities may base a corrective or disciplinary action
   D. other investigators with information so that they may continue with other phases of the investigation

11. In survey work, questionnaires rather than interviews are sometimes used. 11.___
The one of the following which is a DISADVANTAGE of the questionnaire method as compared with the interview is the
   A. difficulty of accurately interpreting the results
   B. problem of maintaining anonymity of the participant
   C. fact that it is relatively uneconomical
   D. requirement of special training for the distribution of questionnaires

12. In his contacts with the public, an employee should attempt to create a good climate of support for his agency. 12.___
This statement is GENERALLY
   A. *false*; such attempts are clearly beyond the scope of his responsibility
   B. *true*; employees of an agency who come in contact with the public have the opportunity to affect public relations
   C. *false*; such activity should be restricted to supervisors trained in public relations techniques
   D. *true*; the future expansion of the agency depends to a great extent on continued public support of the agency

13. The repeated use by a supervisor of a call for volunteers to get a job done is objectionable MAINLY because it 13.___
   A. may create a feeling of animosity between the volunteers and the non-volunteers
   B. may indicate that the supervisor is avoiding responsibility for making assignments which will be most productive

C. is an indication that the supervisor is not familiar with the individual capabilities of his men
D. is unfair to men who, for valid reasons, do not, or cannot volunteer

14. Of the following statements concerning subordinates' expressions to a supervisor of their opinions and feelings concerning work situations, the one which is MOST correct is that 14.
A. by listening and responding to such expressions the supervisor encourages the development of complaints
B. the lack of such expressions should indicate to the supervisor that there is a high level of job satisfaction
C. the more the supervisor listens to and responds to such expressions, the more he demonstrates lack of supervisory ability
D. by listening and responding to such expressions, the supervisor will enable many subordinates to understand and solve their own problems on the job

15. In attempting to motivate employees, rewards are considered preferable to punishment PRIMARILY because 15.
A. punishment seldom has any effect on human behavior
B. punishment usually results in decreased production
C. supervisors find it difficult to punish
D. rewards are more likely to result in willing cooperation

16. In an attempt to combat the low morale in his organization, a high-level supervisor publicized an *open-door* policy to allow employees who wished to do so to come to him with their complaints. 16.
Which of the following is LEAST likely to account for the fact that no employee came in with a complaint?
A. Employees are generally reluctant to go over the heads of their immediate supervisors.
B. The employees did not feel that management would help them.
C. The low morale was not due to complaints associated with the job.
D. The employees felt that they had more to lose than to gain.

17. It is MOST desirable to use written instructions rather than oral instructions for a particular job when 17.
A. a mistake on the job will not be serious
B. the job can be completed in a short time
C. there is no need to explain the job minutely
D. the job involves many details

18. If you receive a telephone call regarding a matter which your office does not handle, you should FIRST 18.___
    A. give the caller the telephone number of the proper office so that he can dial again
    B. offer to transfer the caller to the proper office
    C. suggest that the caller re-dial since he probably dialed incorrectly
    D. tell the caller he has reached the wrong office and then hang up

19. When you answer the telephone, the MOST important reason for identifying yourself and your organization is to 19.___
    A. give the caller time to collect his or her thoughts
    B. impress the caller with your courtesy
    C. inform the caller that he or she has reached the right number
    D. set a business-like tone at the beginning of the conversation

20. As soon as you pick up the phone, a very angry caller begins immediately to complain about city agencies and *red tape*. He says that he has been shifted to two or three different offices. It turns out that he is seeking information which is not immediately available to you. You believe you know, however, where it can be found. Which of the following actions is the BEST one for you to take? 20.___
    A. To eliminate all confusion, suggest that the caller write the agency stating explicitly what he wants.
    B. Apologize by telling the caller how busy city agencies now are, but also tell him directly that you do not have the information he needs.
    C. Ask for the caller's telephone number and assure him you will call back after you have checked further.
    D. Give the caller the name and telephone number of the person who might be able to help, but explain that you are not positive he will get results.

21. Which of the following approaches usually provides the BEST communication in the objectives and values of a new program which is to be introduced? 21.___
    A. A general written description of the program by the program manager for review by those who share responsibility
    B. An effective verbal presentation by the program manager to those affected
    C. Development of the plan and operational approach in carrying out the program by the program manager assisted by his key subordinates
    D. Development of the plan by the program manager's supervisor

22. What is the BEST approach for introducing change? 22.
A
A. combination of written and also verbal communication to all personnel affected by the change
B. general bulletin to all personnel
C. meeting pointing out all the values of the new approach
D. written directive to key personnel

23. Of the following, committees are BEST used for 23.
A. advising the head of the organization
B. improving functional work
C. making executive decisions
D. making specific planning decisions

24. An effective discussion leader is one who 24.
A. announces the problem and his preconceived solution at the start of the discussion
B. guides and directs the discussion according to pre-arranged outline
C. interrupts or corrects confused participants to save time
D. permits anyone to say anything at anytime

25. The human relations movement in management theory is BASICALLY concerned with 25.
A. counteracting employee unrest
B. eliminating the *time and motion* man
C. interrelationships among individuals in organizations
D. the psychology of the worker

KEY (CORRECT ANSWERS)

1. C
2. C
3. A
4. D
5. D

6. B
7. A
8. C
9. A
10. B

11. A
12. B
13. B
14. D
15. D

16. C
17. D
18. B
19. C
20. C

21. C
22. A
23. A
24. B
25. C

REPORT WRITING
EXAMINATION SECTION
TEST 1

DIRECTIONS: Each question or incomplete statement is followed by several suggested answers or completions. Select the one that BEST answers the question or completes the statement. *PRINT THE LETTER OF THE CORRECT ANSWER IN THE SPACE AT THE RIGHT.*

Questions 1-5.

DIRECTIONS: Questions 1 through 5 are to be answered SOLELY on the basis of the following report.

REPORT OF DEFECTIVE EQUIPMENT

DEPARTMENT: *Social Services*
DIVISION: *Personnel*
ROOM: *120B*

REPORT NO. *3026*
DATE OF REPORT: *5/27*

DEFECTIVE EQUIPMENT: *Six office telephones with pick-up and hold buttons*

NUMBERS OF DEFECTIVE TELEPHONES: *533-0102, 3, 4, and 5*

DESCRIPTION OF DEFECT: *Marjorie Black, a Clerk, called on 5/22 to report that the button lights for the four lines on all six telephones in her office were not functioning and it was, therefore, impossible to know which lines were in use. On 5/26, Howard Perl, Admin. Asst., called in regard to the same telephones. He was annoyed because no repairs had been made and stated that all the employees in his unit were being inconvenienced. He requested prompt repair service.*

Ruth Gomez
SIGNATURE OF REPORTING EMPLOYEE
Sr. Telephone Operator
TITLE

JUDITH O'LAUGHLIN
SIGNATURE OF SUPERVISOR

TO BE COMPLETED AFTER SERVICING
DATE: *5/28*
APPROVED: *Judy O'Laughlin*

1. The person who made a written report about the improper functioning of telephones in the Personnel division is 1.___
   A. Marjorie Black
   B. Ruth Gomez
   C. Howard Perl
   D. Judith O'Laughlin

2. How many days elapsed between the original request for telephone repair service and the completion of service? 2.___
   A. 2
   B. 4
   C. 5
   D. 6

3. Of the following, the only information NOT given in the report is 3.___
   A. number of employees affected by the defective service
   B. number of the report
   C. number of telephones with a button defect
   D. telephone numbers of the defective phones

4. The one of the following items of information which would have been LEAST helpful to the repairman who was assigned this repair job is that 4.___
   A. the defect involved pick-up buttons for 4 serviced lines
   B. the location is Room 120B in the Department of Social Services
   C. Marjorie Black initially reported the defective equipment
   D. six telephone units need to be repaired

5. Which of the following statements is CORRECT concerning the people mentioned in the report? 5.___
   A. Ruth Gomez has a higher title than Judith O'Laughlin.
   B. Judith O'Laughlin's signature appears twice on this form.
   C. Howard Perl reported on May 25 that the telephones needed adjusting.
   D. Marjorie Black reported that she was disturbed that no repairs had been made.

Questions 6-10.

DIRECTIONS: Questions 6 through 10 are based on the UNUSUAL OCCURRENC REPORT given below. Five phrases in the report have been removed and are listed below the report as 1. through 5. In each of the five places where phrases of the report have been left out, the number of a question has been inserted. For each question, select the number of the missing phrase which would make the report read correctly

UNUSUAL OCCURRENCE REPORT

POST ____________
TOUR ____________
DATE ____________

Location of Occurrence: ____________

REMARKS: While making rounds this morning, I thought that I heard some strange sounds coming from Storeroom #55. Upon investigation, I saw that __6__ and that the door to the storeroom was slightly opened. At 2:45 A.M., I __7__.

Suddenly two men jumped out from __8__, dropped the tools which they were holding, and made a dash for the door. I ordered them to stop, but they just kept running.

I was able to get a good look at both of them. One man was wearing a green jacket and had a full beard, and the other was short and had blond hair. Immediately, I called the police; and about two minutes later, I notified __9__. I __10__ the police arrived, and I gave them the complete details of the incident.

Security Officer Donald Rimson 2380
Signature Pass No

1. the special inspection control desk
2. behind some crates
3. the lock had been tampered with
4. remained at the storeroom until
5. entered the storeroom and began to look around

6.	A. 1	B. 3	C. 4	D. 5	6.___
7.	A. 2	B. 3	C. 4	D. 5	7.___
8.	A. 1	B. 2	C. 3	D. 4	8.___
9.	A. 1	B. 2	C. 3	D. 4	9.___
10.	A. 2	B. 3	C. 4	D. 5	10.___

Questions 11-13.

DIRECTIONS: Below is a report consisting of 15 numbered sentences, some of which are not consistent with the principles of good report writing. Questions 11 through 13 are to be answered SOLELY on the basis of the information contained in the report and your knowledge of investigative principles and practices.

To: Tom Smith, Administrative Investigator
From: John Jones, Supervising Investigator

1. On January 7, I received a call from Mrs. H. Harris of 684 Sunset Street, Brooklyn.
2. Mrs. Harris informed me that she wanted to report an instance of fraud relating to public assistance payments being received by her neighbor, Mrs. I Wallace.
3. I advised her that such a subject would best be discussed in person.
4. I then arranged a field visitation for January 10 at Mrs. Harris' apartment, 684 Sunset Street, Brooklyn.
5. On January 10, I discussed the basis for Mrs. Harris' charge against Mrs. Wallace at the former's apartment.
6. She stated that her neighbor is receiving Aid to Dependent Children payments for seven children, but that only three of her children are still living with her.
7. In addition, Mrs. Harris also claimed that her husband, whom she reported to the authorities as missing, usually sees her several times a week.
8. After further questioning, Mrs. Harris admitted to me that she had been quite friendly with Mrs. Wallace until they recently argued about trash left in their adjoining hall corridor.
9. However, she firmly stated that her allegations against Mrs. Wallace were valid and that she feared repercussions for her actions.
10. At the completion of the interview, I assured Mrs. Harris of the confidentiality of her statements and that an attempt would be made to verify her allegations.

11. As I was leaving Mrs. Harris' apartment, I noticed a man, aged approximately 45, walking out of Mrs. Wallace's apartment.
12. I followed him until he entered a late model green Oldsmobile and sped away.
13. On January 3, I returned to 684 Sunset Court, having determined that Mrs. Wallace is receiving assistance as indicated by Mrs. Harris.
14. However, upon presentation of official identification Mrs. Wallace refused to admit me to her apartment or grant an interview.
15. I am therefore referring this matter to you for further instructions.

John Jones
Supervising Investigator

11. The one of the following statements that clearly lacks vital information is Statement 11.___
    A. 8 B. 10 C. 12 D. 14

12. Which of the following sentences from the report is ambiguous? 12.___
    Sentence
    A. 2 B. 3 C. 7 D. 10

13. Which of the following sentences contains information contradicting other data in the above report? 13.___
    Sentence
    A. 3 B. 8 C. 10 D. 13

Questions 14-16.

DIRECTIONS: Questions 14 through 16 are to be answered on the basis of the following report.

To: Ralph King
Senior Menagerie Keeper

Date: April 3, 1990

Subject:

From: William Rattner
Menagerie Keeper

This memorandum is to inform you of the disappearance of the boa constrictor from the Reptile Collection in the Main Building.

This morning upon entering the room, I realized that the snake was missing. After having asked around, I am of the opinion that the boa constrictor has been stolen. Since there are no signs of forced entry, it seems likely that whoever removed the snake from the premise entered the room through a window which had been left unlocked the previous night. I, therefore, suggest that all zoo personnel be more concerned with proper security measures in the future so that something like this does not happen again.

14. Which one of the following pieces of information has been OMITTED from the report by the Menagerie Keeper? 14.___
   A. Action taken by him after his discovery that the boa constrictor was missing
   B. The date that the disappearance of the boa constrictor was noted
   C. The time that the disappearance of the boa constrictor was noted
   D. The building in which the boa constrictor was kept

15. Based upon information contained in the above paragraph, which of the following statements would be BEST as the subject of this report? 15.___
   A. Request for more effective security measures in the zoo
   B. Vandalism in the zoo
   C. Disappearance of boa constrictor
   D. Request for replacement of boa constrictor

16. According to the above report, which of the following statements CANNOT be considered factual? 16.___
   A. The boa constrictor was being kept in the Main Building.
   B. The boa constrictor is missing.
   C. All zoo personnel are careless about security measures.
   D. There are no signs of forced entry.

Questions 17-19.

DIRECTIONS: Questions 17 through 19 are to be answered on the basis of the Accident Report below. Read this report carefully before answering the questions. Select your answers ONLY on the basis of this report.

ACCIDENT REPORT

Feb. 14, 1990

On February 14, 1990 at 3:45 P.M., Mr. Warren, while on the top of a stairway at the 34th Street Station, realized the *D* train was in the station loading passengers. In his haste to catch the train, he forcefully ran down the stairs, pushing aside three other people also going down the stairs. Mr. Parker, one of the three people, lost his footing and fell to the bottom of the stairs. Working on the platform, I saw Mr. Parker lose his footing as a result of Mr. Warren's actions, and I immediately went to his aid. Assistant Station Supervisor Brown was attracted to the incident after a crowd had gathered. After 15 minutes, the injured man, Mr. Parker, got up and boarded a train that was in the station and, therefore, he was not hurt seriously.

R. Sands #3214
Conductor

17. Since accident reports should only contain facts, which of the following should NOT be put into the accident report? 17.___
   A. The incident took place at the 34th Street Station.
   B. Mr. Parker was not hurt seriously.
   C. The date that the report was written
   D. Mr. Sands went to the aid of the injured man

18. The title of the person submitting the report was 18.___
   A. Porter
   B. Assistant Station Supervisor
   C. Conductor
   D. Passenger

19. The TOTAL number of different persons mentioned in this report is 19.___
   A. seven B. six C. five D. four

Questions 20-24.

DIRECTIONS: Questions 20 through 24 are to be answered SOLELY on the basis of the following report which is similar to those used in departments for reporting accidents.

REPORT OF ACCIDENT

Date of Accident: 3/21 Time: 3:43 P.M. Date of Report: 3/24

Department Vehicle
Operator's Name: James Doe
Title: Motor Vehicle Operator
Vehicle Code No.: 22-187
License Plate No.: 3N-1234

Damage to Vehicle: Right rear fender ripped, hubcap dented, rear bumper twisted

Vehicle No. 2
Operator's Name: Richard Roe
Operator's Address: 841 W. 68th St.
Owner's Name: Jane Roe
Owner's Address: 2792 Beal Ave.
License Plate No.: 8Y-6789

Damage to Vehicle: Grill, radiator, right side of front bumper, right front fender and headlight crushe

Place of Accident: 8th Avenue & 48th Street

Description of Accident: I was driving east on 48th Street with the green light. I was almost across 8th Avenue when Ford panel truck started north and crashed into my rear right fender. Driver of Ford used abussive language and accused me of rolling into his truck

Persons Injured

Name		Address	
Name	Richard Roe	Address	841 W. 68th Street
Name		Address	
Name		Address	

Witnesses

Name	Richard Roe	Address	341 W. 68th Street
Name	John Brown	Address	226 South Avenue
Name	Mary Green	Address	42 East Street

Report Prepared By James Doe
Title MVO Badge No. 11346

20. According to the above description of the accident, the diagram that would BEST show how and where the vehicles crashed is 20.___

A.

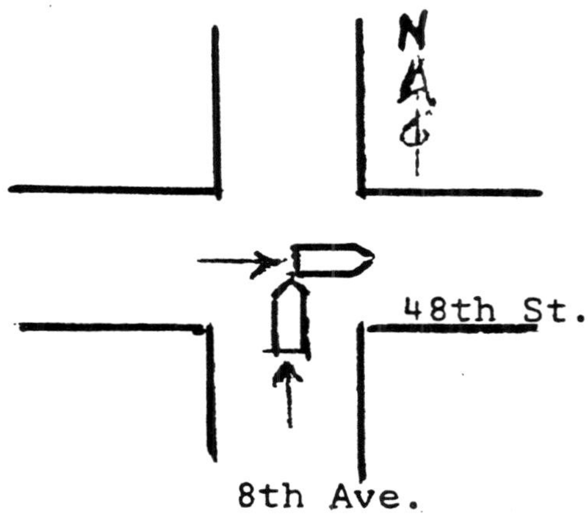

B.

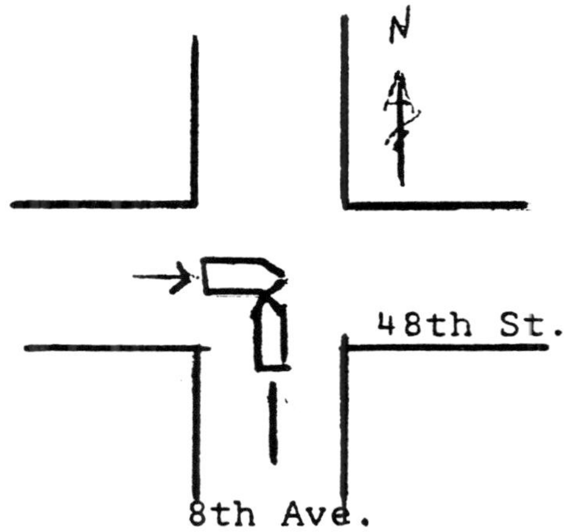

C.

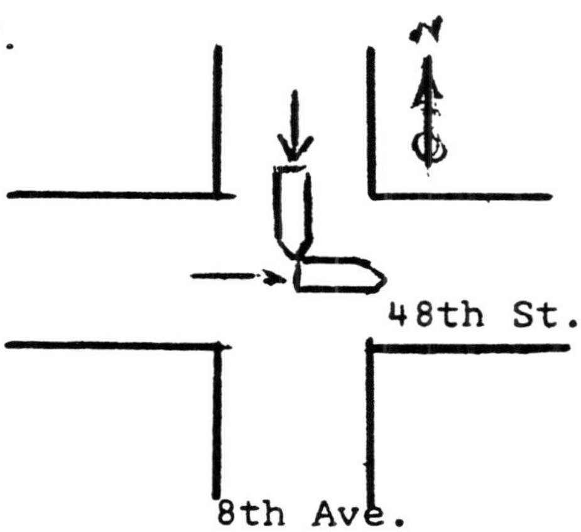

D.

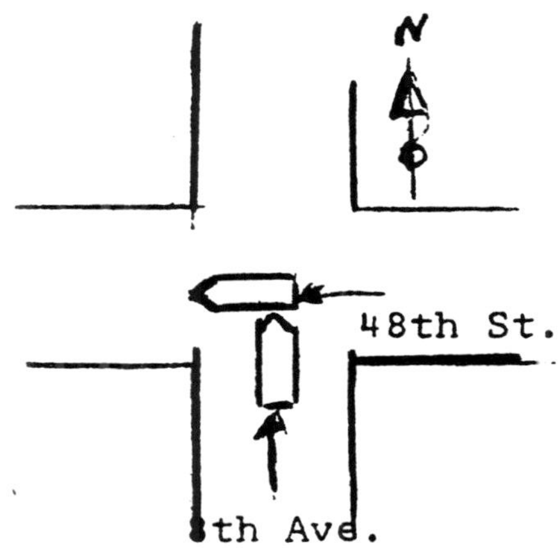

21. Of the following words used in the report, the one spelled INCORRECTLY is 21.___
A. abussive B. accused C. radiator D. twisted

22. The city vehicle involved in this accident can BEST be identified 22.___
   A. as a panel truck
   B. the Department vehicle
   C. by the Badge Number of the operator
   D. by the Vehicle Code Number

23. According to the information in the report, the right-of-way belonged to 23.___
   A. neither vehicle
   B. the Department vehicle
   C. the vehicle that took it
   D. Vehicle No. 2

24. An entry on the report that seems to be INCORRECT is the 24.___
   A. first witness
   B. second witness
   C. third witness
   D. owner's name

25. Assume that the following passage is taken from a report which you, a deputy chief, receive from a battalion chief under your command. The report relates to a fire for which the department received public criticism because of delay in response and extension of fire to neighboring buildings. 25.___

Alarm from box ____ was received at 5:13 P.M. on Friday, October 2. All first alarm companies departed from quarters expeditiously but progress along the vehicle-glutted arterial thoroughfares was agonizingly slow. By dint of extraordinary effort and by virtue of great skill in maneuvering through impassable traffic, Engine Co. ____ arrived at the scene at 5:21 P.M. The sight which greeted them was a virtual Dante's INFERNO, of holocaust proportions. The hub of the conflagration was the penultimate structure of a row of houses, with extension impending to contiguous edifices...

The MAIN fault with the above report is that it
   A. contains spelling and punctuation errors
   B. contains unnecessary details
   C. uses words not in accordance with dictionary definitions
   D. uses inappropriate language and style

KEY (CORRECT ANSWERS)

1. B	6. B	11. C	16. C	21. A
2. D	7. D	12. C	17. B	22. D
3. A	8. B	13. D	18. C	23. B
4. C	9. A	14. C	19. B	24. A
5. B	10. C	15. C	20. A	25. D

TEST 2

DIRECTIONS: Each question or incomplete statement is followed by several suggested answers or completions. Select the one that BEST answers the question or completes the statement. *PRINT THE LETTER OF THE CORRECT ANSWER IN THE SPACE AT THE RIGHT.*

Questions 1-4.

DIRECTIONS: Questions 1 through 4 are to be answered on the basis of the information in the report below.

To: Chief, Division X
From: Mrs. Helen Jones, Clerk
Subject: Accident involving two employees, Mr. John Smith and Mr. Robert Brown

On February 15, 1990, Mr. Smith and Mr. Brown were injured in an accident occurring in the shop at 10 Long Road. No one was in the area of the accident other than Mr. Smith and Mr. Brown. Both of these employees described the following circumstances:

1. Mr. Brown saw the largest tool on the wall begin to fall from where it was hanging and ran up to push Mr. Smith out of the way and to prevent the tool from falling, if possible.
2. Mr. Smith was standing near the wall under some tools which were hanging on nails in the wall.
3. Mr. Brown was standing a few steps from the wall.
4. Mr. Brown stepped toward Mr. Smith, who was on the floor and away from the falling tool. He tripped and fell over a piece of equipment on the floor.
5. Mr. Brown pushed Mr. Smith, who slipped on some grease on the floor and fell to the side, out of the way of the falling tool.
6. Mr. Brown tried to avoid Mr. Smith as he fell. In so doing, he fell against some pipes which were leaning against the wall. The pipes fell on both Mr. Brown and Mr. Smith

Mr. Smith and Mr. Brown were both badly bruised and shaken. They were sent to the General Hospital to determine if any bones were broken. The office was later notified that neither employee was seriously hurt.

Since the accident, matters relating to safety and accident prevention around the shop have occupied the staff. There have been a number of complaints about the location of tools and equipment. Several employees are reluctant to work in the shop unless conditions are improved. Please advise as to the best way to handle this situation.

1. The one of the following which it is MOST important to add to the above memorandum is 1.___
   A. a signature line
   B. a transmittal note
   C. the date of the memo
   D. the initials of the typist

2. The MOST logical order in which to list the circumstances relative to the accident is 2.___
   A. as shown (1, 2, 3, 4, 5, 6)
   B. 2, 3, 1, 5, 4, 6
   C. 1, 5, 4, 6, 3, 2
   D. 3, 2, 4, 6, 1, 5

3. The one of the following which does NOT properly belong with the rest of the memorandum is 3.___
   A. the first section of paragraph 1
   B. the list of circumstances
   C. paragraph 2
   D. paragraph 3

4. According to the information in the memorandum, the BEST description of the subject is: 4.___
   A. Effect of accident on work output of the division
   B. Description of accident involving Mr. Smith and Mr. Brown
   C. Recommendations on how to avoid future accidents
   D. Safety and accident control in the shop

Questions 5-10.

DIRECTIONS: A ferry terminal supervisor is asked to write a report on the incident described in the following passage. Questions 5 through 10 are to be answered on the basis of the incident and the supervisor's report. Your answers should be based on the assumption that everything described in the passage is true.

On July 27, 1990, a rainy, foggy day, Joseph Jones and Steven Smith were in the Whitehall Ferry Terminal at about 9:50 A.M. waiting for the 10:00 A.M. ferry to Staten Island. Smith, seated with his legs stretched out in the aisle, was reading the sports page of the DAILY NEWS. Jones was walking by, drinking ginger ale from a cup. Neither man paid any attention to the other until Jones tripped over Smith's foot, fell to the floor, and dropped his drink. Smith looked at Jones as he lay on the floor and burst out laughing. Jones, infuriated, got up and punched Smith in the jaw. The force of the blow drove Smith's head back against the bench on which he was sitting. Smith did not fight back; he appeared to be dazed. Bystanders called a terminal worker, who assisted in making Smith as comfortable as possible.

One of the other people waiting in the terminal for the ferry was a nurse, who examined Smith and told the ferry terminal supervisor that Smith probably had a concussion. An ambulance was called to take Smith to the hospital. A policeman arrived on the scene.

Jones' injury consisted of a sprained ankle and some bruises, but he refused medical attention. Jones explained to the supervisor what had happened. Jones truly regretted what he had done and went to the local police station with the policeman.

5. Of the following facts about the above incident, which one would be MOST important to include in the ferry terminal supervisor's report? 5.___
   A. The time the next boat was due to arrive
   B. Jones was carrying a cup of ginger ale
   C. Smith was sitting with his legs stretched out in the aisle
   D. Why Smith and Jones were in the terminal

6. The MAIN purpose of writing a report of the above incident is to 6.___
   A. make recommendations for preventing fights in the terminal
   B. state the important facts of the incident
   C. blame Jones for not looking where he was going
   D. provide evidence that Smith was not at fault

7. An adequate report of the above incident MUST give the names of the participants, the names of witnesses, and the 7.___
   A. date, the place, the time, and the events that took place
   B. date, the events that took place, the time, and the names of the terminal personnel on duty that day
   C. place, the names of the terminal personnel on duty that day, the weather conditions, and the events which took place
   D. names of the passengers in the terminal, the time, the place, and the events which took place

8. The supervisor asked for individuals who had witnessed the entire incident to give their account of what they had seen. Thomas White, a twelve-year-old boy said that Jones fell, got up, turned, and then hit Smith. Thomas White's description of the incident is 8.___
   A. *adequate*; it is truthful, straight-forward, and includes necessary details
   B. *adequate*; it shows that the incident was not started on purpose
   C. *inadequate*; he is too young to understand the implications of his testimony
   D. *inadequate*; it omits certain pertinent facts about the incident

9. Another witness, Mary Collins, told the ferry terminal supervisor that when she heard Jones fall, she looked in that direction and saw Jones get up and hit Smith, who was laughing. She immediately ran to find a terminal worker to prevent further fighting. When she returned, she found Smith slumped on the bench. 9.___
Mrs. Collins' report is USEFUL because
   A. it proves that Smith antagonized Jones
   B. it indicates that Jones beat Smith repeatedly
   C. she witnessed that Jones hit Smith
   D. it shows that only one punch was thrown

10. Based on the description given above, which of the following would be the MOST accurate summary for the ferry terminal supervisor to put in his report? 10.___
   A. Jones fell and Smith laughed, which caused Jones to beat him until bystanders got a terminal worker to separate them.
   B. Smith was reading a newspaper when Jones fell. Then Jones hit Smith and dazed him. Smith was examined by a nurse who said that Smith had a serious concussion.
   C. Jones tripped accidentally over Smith's legs and fell. Smith laughed at Jones, who lost his temper and hit Smith, driving Smith's head against the back of a bench.
   D. Smith antagonized Jones first, by tripping him second, by laughing at him, and third by not fighting back. Smith was aided by a nurse and went to the hospital.

Questions 11-13.

DIRECTIONS: Questions 11 through 13 are to be answered SOLELY on the basis of the following report.

To: John Greene, General Park Foreman — Date: May 5

From: Earl Jones, Gardener — Subject:

On May 3rd, as I was finishing a job six feet from the boathouse, I observed that the hole which had been filled in last week was now not level with the ground around it. This seems to be a hazardous condition because it might cause pedestrians to fall into it. I, therefore, suggest that this job be redone as soon as possible.

11. This report should be considered poorly written MAINLY because 11.___
   A. it does not give enough information to take appropriate action
   B. too many different tenses are used

C. it describes no actual personal injury to anyone
D. there is no recommendation in the report to remedy the situation

12. It is noted that the subject of the report has been left out. 12.___
Which of the following statements would be BEST as the subject of this report?
A. Observation made by Earl Jones, Gardener
B. Deteriorating condition of park grounds
C. Report of dangerous condition near boathouse
D. A dangerous walk through the park

13. In order for John Greene to take appropriate action, additional information should be added to the report giving the 13.___
A. exact date the repair was made
B. exact location of the hole
C. exact time the observation was made
D. names of the crew who previously filled in the hole

Questions 14-18.

DIRECTIONS: Questions 14 through 18 consist of sets of four sentences lettered A, B, C, and D. For each question, choose the sentence which is grammatically and stylistically MOST appropriate for use in a formal written report.

14. 14.___
A. It is recommended, therefore, that the impasse panel hearings are to be convened on September 30.
B. It is therefore recommended that the impasse panel hearings be convened on September 30.
C. Therefore, it is recommended to convene the impasse panel hearings on September 30.
D. It is recommended that the impasse panel hearings therefore should be convened on September 30.

15. 15.___
A. Penalties have been assessed for violating the Taylor Law by several unions.
B. When they violated provisions of the Taylor Law, several unions were later penalized.
C. Several unions have been penalized for violating provisions of the Taylor Law.
D. Several unions' violating provisions of the Taylor Law resulted in them being penalized.

16. 16.___
A. The number of disputes settled through mediation has increased significantly over the past two years.
B. The number of disputes settled through mediation are increasing significantly over two-year periods.
C. Over the past two years, through mediation, the number of disputes settled increased significantly.
D. There is a significant increase over the past two years of the number of disputes settled through mediation.

17. A. The union members will vote to determine if the contract is to be approved. 17.___
B. It is not yet known whether the union members will ratify the proposed contract.
C. When the union members vote, that will determine the new contract.
D. Whether the union members will ratify the proposed contract, it is not yet known.

18. A. The parties agreed to an increase in fringe benefits in return for greater worker productivity. 18.___
B. Greater productivity was agreed to be provided in return for increased fringe benefits.
C. Productivity and fringe benefits are interrelated; the higher the former, the more the latter grows.
D. The contract now provides that the amount of fringe benefits will depend upon the level of output by the workers.

19. Of the following excerpts, selected from letters, the one which is considered by modern letter writing experts to be the BEST is: 19.___
A. Attached please find the application form to be filled out by you. Return the form to this office at the above address.
B. Forward to this office your check accompanied by the application form enclosed with this letter.
C. If you wish to apply, please complete and return the enclosed form with your check.
D. In reply to your letter of December ___, enclosed herewith please find the application form you requested.

20. A city employee who writes a letter requesting information from a businessman should realize that, of the following, it is MOST important to 20.___
A. end the letter with a polite closing
B. make the letter short enough to fit on one page
C. use a form, such as a questionnaire, to save the businessman's time
D. use a courteous tone that will get the desired cooperation

Questions 21-22.

DIRECTIONS: Questions 21 and 22 consist of four sentences. Choose the one sentence in each set of four that would be BEST for a formal letter or report. Consider grammar and appropriate usage.

21. A. Most all the work he completed before he become ill. 21.___
B. He completed most of the work before becoming ill.
C. Prior to him becoming ill his work was mostly completed.
D. Before he became ill most of the work he had completed.

22. A. Being that the report lacked a clearly worded recommendation, it did not matter that it contained enough information. 22.___
    B. There was enough information in the report, although it, including the recommendation, were not clearly worded.
    C. Although the report contained enough information, it did not have a clearly worded recommendation.
    D. Though the report did not have a recommendation that was clearly worded, and the information therein contained was enough.

Questions 23-25.

DIRECTIONS: In Questions 23 through 25, choose the sentence which is BEST from the point of view of English usage suitable for a business letter or report.

23. A. Answering of veterans' inquiries, together with the receipt of fees, have been handled by the Bursar's Office since the new President came. 23.___
    B. Since the new President's arrival, the handling of all veteran's inquiries has been turned over to the Bursar's Office.
    C. In addition to the receipt of fees, the Bursar's Office has been handling veterans' inquiries since the new President came.
    D. The principle change in the work of the Bursar's Office since the new President came is that it now handles veterans' inquiries as well as the receipt of fees.

24. A. The current unrest about education undoubtedly stems in part from the fact that the people fear the basic purposes of the schools are being neglected or supplanted by spurious ones. 24.___
    B. The fears of people that the basic purposes of the schools are being neglected or supplanted by spurious ones contributes to the current unrest about education.
    C. Undoubtedly some responsibility for the current unrest about education must be assigned to peoples' fears that the purpose and base of the school system is being neglected or supplanted.
    D. From the fears of people that the basic purposes of the schools are being neglected or supplanted by spurious ones undoubtedly stem in part the current unrest about education.

25. A. The existence of administrative phenomena are clearly established, but their characteristics, relations and laws are obscure. 25.___
    B. The obscurity of the characteristics, relations and laws of administrative phenomena do not preclude their existence.

C. Administrative phenomena clearly exists in spite of the obscurity of their characteristics, relations and laws.
D. The characteristics, relations and laws of administrative phenomena are obscure but the existence of the phenomena is clear.

KEY (CORRECT ANSWERS)

1. C
2. B
3. D
4. B
5. C

6. B
7. A
8. D
9. C
10. C

11. A
12. C
13. B
14. B
15. C

16. A
17. B
18. A
19. C
20. D

21. B
22. C
23. C
24. A
25. D

TEST 3

DIRECTIONS: Each question or incomplete statement is followed by several suggested answers or completions. Select the one that BEST answers the question or completes the statement. *PRINT THE LETTER OF THE CORRECT ANSWER IN THE SPACE AT THE RIGHT.*

1. Of the following, the BEST statement concerning the placement of *Conclusions and Recommendations* in a management report is: 1.___
   A. Recommendations should always be included in a report unless the report presents the results of an investigation.
   B. If a report presents conclusions, it must present recommendations.
   C. Every statement that is a conclusion should grow out of facts given elsewhere in the report.
   D. Conclusions and recommendations should always conclude the report because they depend on its contents.

2. Assume you are preparing a systematic analysis of your agency's pest control program and its effect on eliminating rodent infestation of premises in a specific region. To omit from your report important facts which you originally received from the person to whom you are reporting is GENERALLY considered to be 2.___
   A. *desirable*; anyone who is likely to read the report can consult his files for extra information
   B. *undesirable*; the report should include major facts that are obtained as a result of your efforts
   C. *desirable*; the person you are reporting to does not pass the report on to others who lack his own familiarity with the subject
   D. *undesirable*; the report should include all of the facts that are obtained as a result of your efforts

3. Of all the non-verbal devices used in report writing, tables are used most frequently to enable a reader to compare statistical information more easily. Hence, it is important that an analyst know when to use tables. Which one of the following statements that relate to tables is generally considered to be LEAST valid? 3.___
   A. A table from an outside source must be acknowledged by the report writer.
   B. A table should be placed far in advance of the point where it is referred to or discussed in the report.
   C. The notes applying to a table are placed at the bottom of the table, rather than at the bottom of the page on which the table is found.
   D. A table should indicate the major factors that effect the data it contains.

4. Assume that an analyst writes reports which contain more detail than might be needed to serve their purpose. 4.___
Such a practice is GENERALLY considered to be
   A. *desirable*; this additional detail permits maximized machine utilization
   B. *undesirable*; if specifications of reports are defined when they are first set up, loss of flexibility will follow
   C. *desirable*; everything ought to be recorded so it will be there if it is ever needed
   D. *undesirable*; recipients of these reports are likely to discredit them entirely

Questions 5-6.

DIRECTIONS: Questions 5 and 6 consist of sentences lettered A, B, C, and D. For each question, choose the sentence which is stylistically and grammatically MOST appropriate for a management report.

5. A. For too long, the citizen has been forced to rely for his productivity information on the whims, impressions and uninformed opinion of public spokesmen. 5.___
   B. For too long, the citizen has been forced to base his information about productivity on the whims, impressions and uninformed opinion of public spokesmen.
   C. The citizen has been forced to base his information about productivity on the whims, impressions and uninformed opinion of public spokesmen for too long.
   D. The citizen has been forced for too long to rely for his productivity information on the whims, impressions and uninformed opinion of public spokesmen.

6. A. More competition means lower costs to the city, thereby helping to compensate for inflation. 6.___
   B. More competition, helping to compensate for inflation, means lower costs to the city.
   C. Inflation may be compensated for by more competition, which will reduce the city's costs.
   D. The costs to the city will be lessened by more competition, helping to compensate for inflation.

Questions 7-11.

DIRECTIONS: In Questions 7 through 11, choose the sentence which is BEST from the point of view of English usage suitable for a business letter or report.

7. A. It is the opinion of the Commissioners that programs which include the construction of cut-rate municipal garages in the central business district is inadvisable. 7.___
   B. Having reviewed the material submitted, the program for putting up cut-rate garages in the central business district seemed likely to cause traffic congestion.

C. The Commissioners believe that putting up cut-rate municipal garages in the central business district is inadvisable.
D. Making an effort to facilitate the cleaning of streets in the central business district, the building of cut-rate municipal garages presents the problem that it would encourage more motorists to come into the central city.

8. A. This letter, together with the reports, are to be sent to the principal. 8.___
B. The reports, together with this letter, is to be sent to the principal.
C. The reports and this letter is to be sent to the principal.
D. This letter, together with the reports, is to be sent to the principal.

9. A. Each employee has to decide for themselves whether to take the examination. 9.___
B. Each of the employees has to decide for himself whether to take the examination.
C. Each of the employees has to decide for themselves whether to take the examination.
D. Each of the employees have to decide for himself whether to take the examination.

10. A. The reason a new schedule is being prepared is that there has been a change in priorities. 10.___
B. Because there has been a change in priorities is the reason why a new schedule is being made up.
C. The reason why a new schedule is being made up is because there has been a change in priorities.
D. Because of a change in priorities is the reason why a new schedule is being prepared.

11. A. The changes in procedure had an unfavorable affect upon the output of the unit. 11.___
B. The increased output of the unit was largely due to the affect of the procedural changes.
C. The changes in procedure had the effect of increasing the output of the unit.
D. The increased output of the unit from the procedural changes were the effect.

Questions 12-19.

DIRECTIONS: Questions 12 through 19 each consist of four sentences. Choose the one sentence in each set of four that would be BEST for a formal letter or report. Consider grammar and appropriate usage.

12. A. These statements can be depended on, for their truth has been guaranteed by reliable city employees. 12.___
    B. Reliable city employees guarantee the facts with regards to the truth of these statements.
    C. Most all these statements have been supported by city employees who are reliable and can be depended upon.
    D. The city employees which have guaranteed these statements are reliable.

13. A. I believe the letter was addressed to either my associate or I. 13.___
    B. If properly addressed, the letter will reach my associate and I.
    C. My associate's name, as well as mine, was on the letter.
    D. The letter had been addressed to myself and my associate.

14. A. The secretary would have corrected the errors if she knew that the supervisor would see the report. 14.___
    B. The supervisor reprimanded the secretary, whom she believed had made careless errors.
    C. Many errors were found in the report which she typed and could not disregard them.
    D. The errors in the typed report were so numerous that they could hardly be overlooked.

15. A. His consultant was as pleased as he with the success of the project. 15.___
    B. The success of the project pleased both his consultant and he.
    C. He and also his consultant was pleased with the success of the project.
    D. Both his consultant and he was pleased with the success of the project.

16. A. Since the letter did not contain the needed information, it was not real useful to him. 16.___
    B. Being that the letter lacked the needed information, he could not use it.
    C. Since the letter lacked the needed information, it was of no use to him.
    D. This letter was useless to him because there was no needed information in it.

17. A. Scarcely had the real estate tax increase been declared than the notices were sent out. 17.___
    B. They had no sooner declared the real estate tax increases when they sent the notices to the owners.
    C. The city had hardly declared the real estate tax increase till the notices were prepared for mailing.
    D. No sooner had the real estate tax increase been declared than the notices were sent out.

18. A. Though deeply effected by the setback, the advice given by the admissions office began to seem more reasonable. 18.___
B. Although he was deeply effected by the setback, the advice given by the admissions office began to seem more reasonable.
C. Though the setback had affected him deeply, the advise given by the admissions office began to seem more reasonable.
D. Although he was deeply affected by the setback, the advice given by the admissions office began to seem more reasonable.

19. A. Returning to the administration building after attendance at a meeting, the door was locked despite an agreement that it would be left open. 19.___
B. When he returned to the administration building after attending a meeting, he found the door locked, despite an agreement that it would be left open.
C. After attending a meeting, the door to the administration building was locked, despite an agreement that it would be left open.
D. When he returned to the administration building after attendance at a meeting, he found the door locked, despite an agreement that it would be left open.

20. A formal business report may consist of many parts, including the following: 20.___
1. Table of Contents
2. List of references
3. Preface
4. Index
5. List of tables
6. Conclusions or recommendations

Of the following, in setting up a formal report, the PROPER order of the six parts listed is:
A. 1, 3, 6, 5, 2, 4
B. 4, 3, 2, 5, 6, 1
C. 3, 1, 5, 6, 2, 4
D. 2, 5, 3, 1, 4, 6

21. Suppose you are writing a report on an interview you have just completed with a particularly hostile applicant for public assistance. 21.___
Which of the following BEST describes what you should include in this report?
A. What you think caused the applicant's hostile attitude during the interview
B. Specific examples of the applicant's hostile remarks and behavior
C. The relevant information uncovered during the interview
D. A recommendation that the applicant's request be denied because of his hostility

22. When including recommendations in a report to your supervisor, which of the following is MOST important for you to do? 22.___
   A. Provide several alternative courses of action for each recommendation.
   B. First present the supporting evidence, then the recommendations.
   C. First present the recommendations, then the supporting evidence.
   D. Make sure the recommendations arise logically out of the information in the report.

23. It is often necessary that the writer of a report present facts and sufficient arguments to gain acceptance of the points, conclusions, or recommendations set forth in the report. 23.___
Of the following, the LEAST advisable step to take in organizing a report, when such argumentation is the important factor, is a(n)
   A. elaborate expression of personal belief
   B. businesslike discussion of the problem as a whole
   C. orderly arrangement of convincing data
   D. reasonable explanation of the primary issues

24. Assume that a clerk is asked to prepare a special report which he has not prepared before. He decides to make a written outline of the report before writing it in full. This decision by the clerk is 24.___
   A. *good*, mainly because it helps the writer to organize his thoughts and decide what will go into the report
   B. *good*, mainly because it clearly shows the number of topics, number of pages, and the length of the report
   C. *poor*, mainly because it wastes the time of the writer since he will have to write the full report anyway
   D. *poor*, mainly because it confines the writer to those areas listed in the outline

25. Assume that a clerk in the water resources central shop is asked to prepare an important report, giving the location and condition of various fire hydrants in the city. One of the hydrants in question is broken and is spewing rusty water in the street, creating a flooded condition in the area. The clerk reports that the hydrant is broken but does not report the escaping water or the flood. 25.___
Of the following, the BEST evaluation of the clerk's decision about what to report is that it is basically
   A. *correct*; chiefly because a lengthy report would contain irrelevant information
   B. *correct*; chiefly because a more detailed description of a hydrant should be made by a fireman, not a clerk
   C. *incorrect*; chiefly because the clerk's assignment was to describe the condition of the hydrant and he should give a full explanation
   D. *incorrect*; chiefly because the clerk should include as much information as possible in his report whether or not it is relevant

KEY (CORRECT ANSWERS)

1. C
2. B
3. B
4. D
5. B

6. A
7. C
8. D
9. B
10. A

11. C
12. A
13. C
14. D
15. A

16. C
17. D
18. D
19. B
20. C

21. C
22. D
23. A
24. A
25. C

REPORT WRITING

EXAMINATION SECTION

DIRECTIONS FOR THIS SECTION:
Each question or incomplete statement is followed by several suggested answers or completions. Select the one that BEST answers the question or completes the statement. *PRINT THE LETTER OF THE CORRECT ANSWER IN THE SPACE AT THE RIGHT.*

TEST 1

Questions 1-4.
DIRECTIONS: Answer Questions 1 through 4 on the basis of the following report which was prepared by a supervisor for inclusion in his agency's annual report.

On Oct. 13, 1991, I was assigned to study the salaries paid to clerical employees in various titles by the city and by private industry in the area.

In order to get the data I needed, I called Mr. Johnson at the Bureau of the Budget and the payroll officers at X Corp. - a brokerage house, Y Co. - an insurance company, and Z Inc. - a publishing firm. None of them was available and I had to call all of them again the next day.

When I finally got the information I needed, I drew up a chart, which is attached. Note that not all of the companies I contacted employed people at all the different levels used in the city service.

The conclusions I draw from analyzing this information is as follows: The city's entry-level salary is about average for the region; middle-level salaries are generally higher in the city government than in private industry; but salaries at the highest levels in private industry are better than city employees' pay.

1. Which of the following criticisms about the style in which this report is written is *most valid?* 1. ...
   A. It is too informal. B. It is too concise.
   C. It is too choppy. D. The syntax is too complex.
2. Judging from the statements made in the report, the method followed by this employee in performing his research was 2. ...
   A. *good;* he contacted a representative sample of businesses in the area
   B. *poor;* he should have drawn more definite conclusions
   C. *good;* he was persistent in collecting information
   D. *poor;* he did not make a thorough study
3. One sentence in this report contains a grammatical error. This sentence *begins* on line number 3. ...
   A. 4 B. 7 C. 10 D. 13
4. The type of information given in this report which should be presented in footnotes or in an appendix, is the 4. ...
   A. purpose of the study
   B. specifics about the businesses contacted
   C. reference to the chart
   D. conclusions drawn by the author
5. The use of a graph to show statistical data in a report is *superior* to a table because it 5. ...
   A. features approximations
   B. emphasizes facts and relationships more dramatically

C. presents data more accurately
D. is easily understood by the average reader

6. Of the following, the degree of formality required of a written report in tone is *most likely* to depend on the 6. ...
A. subject matter of the report
B. frequency of its occurrence
C. amount of time available for its preparation
D. audience for whom the report is intended

7. Of the following, a distinguishing characteristic of a written report intended for the head of your agency as compared to a report prepared for a lower-echelon staff member, is that the report for the agency head should *usually* include 7. ...
A. considerably more detail, especially statistical data
B. the essential details in an abbreviated form
C. all available source material
D. an annotated bibliography

8. Assume that you are asked to write a lengthy report for use by the administrator of your agency, the subject of which is "The Impact of Proposed New Data Processing Operations on Line Personnel" in your agency. You decide that the *most appropriate* type of report for you to prepare is an analytical report, including recommendations. 8. ...
The MAIN reason for your decision is that
A. the subject of the report is extremely complex
B. large sums of money are involved
C. the report is being prepared for the administrator
D. you intend to include charts and graphs

9. Assume that you are preparing a report based on a survey dealing with the attitudes of employees in Division X regarding proposed new changes in compensating employees for working overtime. Three per cent of the respondents to the survey voluntarily offer an unfavorable opinion on the method of assigning overtime work, a question not specifically asked of the employees. 9. ...
On the basis of this information, the *most appropriate* and *significant* of the following comments for you to make in the report with regard to employees' attitudes on assigning overtime work, is that
A. an insignificant percentage of employees dislike the method of assigning overtime work
B. three per cent of the employees in Division X dislike the method of assigning overtime work
C. three per cent of the sample selected for the survey voiced an unfavorable opinion on the method of assigning overtime work
D. some employees voluntarily voiced negative feelings about the method of assigning overtime work, making it impossible to determine the extent of this attitude

10. A supervisor should be able to prepare a report that is well-written and unambiguous. 10. ...
Of the following sentences that might appear in a report, select the one which communicates *most clearly* the intent of its author.
A. When your subordinates speak to a group of people, they should be well-informed.

B. When he asked him to leave, SanMan King told him that he would refuse the request.
C. Because he is a good worker, Foreman Jefferson assigned Assistant Foreman D'Agostino to replace him.
D. Each of us is responsible for the actions of our subordinates.

11. In some reports, especially longer ones, a list of the resources (books, papers, magazines, etc.) used to prepare it is included. This list is called the 11. ...
A. accreditation B. bibliography
C. summary D. glossary

12. Reports are usually divided into several sections, some of which are more necessary than others. 12. ...
Of the following, the section which is ABSOLUTELY necessary to include in a report is
A. a table of contents B. the body
C. an index D. a bibliography

13. Suppose you are writing a report on an interview you have just completed with a particularly hostile applicant. 13. ...
Which of the following BEST describes what you should include in this report?
A. What you think caused the applicant's hostile attitude during the interview
B. Specific examples of the applicant's hostile remarks and behavior
C. The relevant information uncovered during the interview
D. A recommendation that the applicant's request be denied because of his hostility

14. When including recommendations in a report to your supervistor, which of the following is MOST important for you to do? 14. ...
A. Provide several alternative courses of action for each recommendation
B. First present the supporting evidence, then the recommendations
C. First present the recommendations, then the supporting evidence
D. Make sure the recommendations arise logically out of the information in the report

15. It is often necessary that the writer of a report present facts and sufficient arguments to gain acceptance of the points, conclusions, or recommendations set forth in the report. 15. ...
Of the following, the LEAST advisable step to take in organizing a report, when such argumentation is the important factor, is a(n)
A. elaborate expression of personal belief
B. businesslike discussion of the problem as a whole
C. orderly arrangement of convincing data
D. reasonable explanation of the primary issues

16. In some types of reports, visual aids add interest, meaning, and support. They also provide an essential means of effectively communicating the message of the report. 16. ...
Of the following, the selection of the suitable visual aids to use with a report is LEAST dependent on the
A. nature and scope of the report

B. way in which the aid is to be used
C. aids used in other reports
D. prospective readers of the report

17. Visual aids used in a report may be placed either in the text material or in the appendix. Deciding where to put a chart, table, or any such aid *should* depend on the 17. ...
A. title of the report B. purpose of the visual aid
C. title of the visual aid D. length of the report

18. A report is often revised several times before final preparation and distribution in an effort to make certain the report meets the needs of the situation for which it is designed. 18. ...
Which of the following is the BEST way for the author to be sure that a report covers the areas he intended?
A. Obtain a co-worker's opinion
B. Compare it with a content checklist
C. Test it on a subordinate D. Check his bibliography

19. In which of the following situations is an oral report preferable to a written report? When a(n) 19. ...
A. recommendation is being made for a future plan of action
B. department head requests immediate information
C. long standing policy change is made
D. analysis of complicated statistical data is involved

20. When an applicant is approved, the supervisor must fill in standard forms with certain information. 20. ...
The GREATEST advantage of using standard forms in this situation rather than having the supervisor write the report as he sees fit, is that
A. the report can be acted on quickly
B. the report can be written without directions from a supervisor
C. needed information is less likely to be left out of the report
D. information that is written up this way is more likely to be verified

21. Assume that it is part of your job to prepare a monthly report for your unit head that eventually goes to the director. The report contains information on the number of applicants you have interviewed that have been approved and the number of applicants you have interviewed that have been turned down. 21. ...
Errors on such reports are SERIOUS because
A. you are expected to be able to prove how many applicants you have interviewed each month
B. accurate statistics are needed for effective management of the department
C. they may not be discovered before the report is transmitted to the director
D. they may result in loss to the applicants left out of the report

22. The frequency with which job reports are submitted should depend MAINLY on 22. ...
A. how comprehensive the report has to be
B. the amount of information in the report
C. the availability of an experienced man to write the report

D. the importance of changes in the information included in the report

23. The CHIEF purpose in preparing an outline for a report is *usually* to insure that 23. ...
A. the report will be grammatically correct
B. every point will be given equal emphasis
C. principal and secondary points will be properly integrated
D. the language of the report will be of the same level and include the same technical terms

24. The MAIN reason for requiring written job reports is to 24. ...
A. avoid the necessity of oral orders
B. develop better methods of doing the work
C. provide a permanent record of what was done
D. increase the amount of work that can be done

25. Assume you are recommending in a report to your supervisor that a radical change in a standard maintenance procedure should be adopted. 25. ...
Of the following, the MOST important information to be included in this report is
A. a list of the reasons for making this change
B. the names of others who favor the change
C. a complete description of the present procedure
D. amount of training time needed for the new procedure

TEST 2

1. It is often necessary that the writer of a report present facts and sufficient arguments to gain acceptance of the points, conclusions, or recommendations set forth in the report. 1. ...
Of the following, the LEAST advisable step to take in organizing a report, when such argumentation is the important factor, is a(n)
A. elaborate expression of personal belief
B. businesslike discussion of the problem as a whole
C. orderly arrangement of convincing data
D. reasonable explanation of the primary issues

2. Of the following, the factor which is generally considered to be LEAST characteristic of a good control report is that it 2. ...
A. stresses performance that adheres to standard rather than emphasizing the exception
B. supplies information intended to serve as the basis for corrective action
C. provides feedback for the planning process
D. includes data that reflect trends as well as current status

3. An administrative assistant has been asked by his superior to write a concise, factual report with objective conclusions and recommendations based on facts assembled by other researchers. 3. ...
Of the following factors, the administrative assistant should give LEAST consideration to

A. the educational level of the person or persons for whom the report is being prepared
B. the use to be made of the report
C. the complexity of the problem
D. his own feelings about the importance of the problem

4. When making a written report, it is often recommended that the findings or conclusions be presented near the beginning of the report. 4. ...
Of the following, the MOST important reason for doing this is that it
A. facilitates organizing the material clearly
B. assures that all the topics will be covered
C. avoids unnecessary repetition of ideas
D. prepares the reader for the facts that will follow

5. You have been asked to write a report on methods of hiring and training new employees. Your report is going to be about ten pages long. 5. ...
For the convenience of your readers, a brief summary of your findings *should*
A. appear at the beginning of your report
B. be appended to the report as a postscript
C. be circulated in a separate memo
D. be inserted in tabular form in the middle of your report

6. In preparing a report, the MAIN reason for writing an outline is *usually* to 6. ...
A. help organize thoughts in a logical sequence
B. provide a guide for the typing of the report
C. allow the ultimate user to review the report in advance
D. ensure that the report is being prepared on schedule

7. The one of the following which is *most appropriate* as a reason for including footnotes in a report is to 7. ...
A. correct capitalization B. delete passages
C. improve punctuation D. cite references

8. A completed formal report may contain all of the following EXCEPT 8. ...
A. a synopsis B. a preface
C. marginal notes D. bibliographical references

9. Of the following, the MAIN use of proofreaders' marks is to 9. ...
A. explain corrections to be made
B. indicate that a manuscript has been read and approved
C. let the reader know who proofread the report
D. indicate the format of the report

10. Informative, readable and concise reports have been found to observe the following rules: 10. ...
Rule I. Keep the report short and easy to understand.
Rule II. Vary the length of sentences.
Rule III. Vary the style of sentences so that, for example, they are not all just subject-verb, subject-verb.
Consider this hospital laboratory report: The experiment was started in January. The apparatus was put together in six weeks. At that time the synthesizing process was begun. The synthetic chemicals were separated. Then they were used in tests on patients.
Which one of the following choices MOST accurately classifies the above rules into those which are *violated* by this report and those which are *not?*

A. II is violated, but I and III are not.
B. III is violated, but I and II are not.
C. II and III are violated, but I is not.
D. I, II, and III are violated.

Questions 11-13.

DIRECTIONS: Questions 11 through 13 are based on the following example of a report. The report consists of eight numbered sentences, some of which are not consistent with the principles of good report writing.

(1) I interviewed Mrs. Loretta Crawford in Room 424 of County Hospital. (2) She had collapsed on the street and been brought into emergency. (3) She is an attractive woman with many friends judging by the cards she had received. (4) She did not know what her husband's last job had been, or what their present income was. (5) The first thing that Mrs. Crawford said was that she had never worked and that her husband was presently unemployed. (6) She did not know if they had any medical coverage or if they could pay the bill. (7) She said that her husband could not be reached by telephone but that he would be in to see her that afternoon. (8) I left word at the nursing station to be called when he arrived.

11. A good report should be arranged in logical order. Which of the following sentences from the report does NOT appear in its proper sequence in the report? Sentence 11. ...
A. 1 B. 4 C. 7 D. 8

12. Only material that is relevant to the main thought of a report should be included. 12. ...
Which of the following sentences from the report contains material which is LEAST relevant to this report? Sentence
A. 3 B. 4 C. 6 D. 8

13. Reports should include all essential information. 13. ...
Of the following, the MOST important fact that is *missing* from this report is:
A. Who was involved in the interview
B. What was discovered at the interview
C. When the interview took place
D. Where the interview took place

Questions 14-15.

DIRECTIONS: Each of Questions 14 and 15 consists of four numbered sentences which constitute a paragraph in a report. They are not in the right order. Choose the numbered arrangement appearing after letter A, B, C, or D which is MOST logical and which BEST expresses the thought of the paragraph.

14. 14. ...
I. Congress made the commitment explicit in the Housing Act of 1949, establishing as a national goal the realization of a decent home and suitable environment for every American family.
II. The result has been that the goal of decent home and suitable environment is still as far distant as ever for the disadvantaged urban family.
III. In spite of this action by Congress, federal housing programs have continued to be fragmented and grossly under-funded.
IV. The passage of the National Housing Act signaled a new federal commitment to provide housing for the nation's citizens.

A. I, IV, III, II B. IV, I, III, II

C. IV, I, III, II D. II, IV, I, III

15. I. The greater expense does not necessarily involve "exploitation," but it is often perceived as exploitative and unfair by those who are aware of the price differences involved, but unaware of operating costs. 15. ...

II. Ghetto residents believe they are "exploited" by local merchants, and evidence substantiates some of these beliefs.

III. However, stores in low-income areas were more likely to be small independents, which could not achieve the economies available to supermarket chains and were, therefore, more likely to charge higher prices, and the customers were more likely to buy smaller-sized packages which are more expensive per unit of measure.

IV. A study conducted in one city showed that distinctly higher prices were charged for goods sold in ghetto stores than in other areas.

A. IV, II, I, III B. IV, I, III, II
C. II, IV, III, I D. II, III, IV, I

16. In organizing data to be presented in a formal report, the FIRST of the following steps should be 16. ...
A. determining the conclusions to be drawn
B. establishing the time sequence of the data
C. sorting and arranging like data into groups
D. evaluating how consistently the data support the recommendations

17. All reports should be prepared with *at least* one copy so that 17. ...
A. there is one copy for your file
B. there is a copy for your supervisor
C. the report can be sent to more than one person
D. the person getting the report can forward a copy to someone else

18. Before turning in a report of an investigation he has made, a supervisor discovers some additional information he did not include in this report. 18. ...
Whether he rewrites this report to include this additional information should PRIMARILY depend on the
A. importance of the report itself
B. number of people who will eventually review this report
C. established policy covering the subject matter of the report
D. bearing this new information has on the conclusions of the report

KEYS (CORRECT ANSWERS)

TEST 1

1. A	9. D	18. B
2. D	10. D	19. B
3. D	11. B	20. C
4. B	12. B	21. B
5. B	13. C	22. D
6. D	14. D	23. C
7. B	15. A	24. C
8. A	16. C	25. A
	17. B	

TEST 2

1. A	10. C
2. A	11. B
3. D	12. A
4. D	13. C
5. A	14. B
6. A	15. C
7. D	16. C
8. C	17. A
9. A	18. D

DOCUMENTS AND FORMS

PREPARING WRITTEN MATERIALS

EXAMINATION SECTION

DIRECTIONS: Each question or incomplete statement is followed by several suggested answers or completions. Select the one that BEST answers the question or completes the statement. *PRINT THE LETTER OF THE CORRECT ANSWER IN THE SPACE AT THE RIGHT.*

TEST 1

1. Of the following forms, the one in which horizontal lines may BEST be omitted is one 1.____
   A. that is to be filled in by hand
   B. that is to be filled in by typewriter
   C. which requires many fill-ins
   D. with little room for fill-ins

2. A certain form letter starts with the words *Dear Mr.* followed by a blank space. 2.____
   The MAJOR shortcoming in this is that
   A. salutations should not be placed on form letters
   B. *Gentlemen:* is preferable in a formal business letter
   C. the name will have to be typed in
   D. this salutation may be inappropriate

3. *Form paragraphs* may BEST be defined as 3.____
   A. block-style paragraphs
   B. paragraphs on a form
   C. paragraphs within a form letter
   D. standardized paragraphs used in correspondence

4. In general, the CHIEF economy of using multicopy forms is in 4.____
   A. the paper on which the form is printed
   B. printing the form
   C. employee time
   D. carbon paper

5. Suppose your supervisor has asked you to develop a form to record certain information needed. 5.____
   The FIRST thing you should do is to
   A. determine the type of data that will be recorded repeatedly so that it can be preprinted
   B. study the relationship of the form to the job to be accomplished so that the form can be planned
   C. determine the information that will be recorded in the same place on each copy of the form so that it can be used as a check
   D. find out who will be responsible for supplying the information so that space can be provided for their signatures

6. Which of the following is MOST likely to reduce the volume of paperwork in a unit responsible for preparing a large number of reports? 6.____
   A. Changing the office layout so that there will be a minimum of backtracking and delay
   B. Acquiring additional adding and calculating machines
   C. Consolidating some of the reports
   D. Inaugurating a *records retention* policy to reduce the length of time office papers are retained

7. Of the following basic guides to effective letter writing, which one would NOT be recommended as a way of improving the quality of business letters? 7.____
   A. Use emphatic phrases like *close proximity* and *first and foremost* to round out sentences.
   B. Break up complicated sentences by making short sentences out of dependent clauses.
   C. Replace old-fashioned phrases like *enclosed please find* and *recent date* with a more direct approach.
   D. Personalize letters by using your reader's name at least once in the body of the message.

8. Suppose that you must write a reply letter to a citizen's request for a certain pamphlet printed by your agency. The pamphlet is temporarily unavailable but a new supply will be arriving by December 8 or 9. 8.____
   Of the following four sentences, which one expresses the MOST positive business letter writing approach?
   A. We cannot send the materials you requested until after December 8.
   B. May we assure you that the materials you requested will be sent as quickly as possible.
   C. We will be sending the materials you requested as soon as our supply is replenished.
   D. We will mail the materials you requested on or shortly after December 8.

9. Using form letters in business correspondence is LEAST effective when 9.____
   A. answering letters on a frequently recurring subject
   B. giving the same information to many addresses
   C. the recipient is only interested in the routine information contained in the form letter
   D. a reply must be keyed to the individual requirements of the intended reader

10. The ability to write memos and letters is very important in clerical and administrative work. Methodical planning of a reply letter usually involves the following basic steps which are arranged in random order: 10.____
   I. Determine the purpose of the letter you are about to write.
   II. Make an outline of what information your reply letter should contain.

III. Read carefully the letter to be answered to find out its main points.
IV. Assemble the facts to be included in your reply letter.
V. Visualize your intended reader and adapt your letter writing style to him.

If the above numbered steps were arranged in their proper logical order, the one which would be THIRD in the sequence is
A. II B. III C. IV D. V

11. Generally, the frequency with which reports are to be submitted or the length of the interval which they cover should depend MAINLY on the 11.____
A. amount of time needed to prepare the reports
B. degree of comprehensiveness required in the reports
C. availability of the data to be included in the reports
D. extent of the variations in the data with the passage of time

12. The objectiveness of a report is its unbiased presentation of the facts. 12.____
If this is so, which of the following reports listed below is likely to be the MOST objective?
A. The Best Use of an Electronic Computer in Department Z
B. The Case for Raising the Salaries of Employees in Department A
C. Quarterly Summary of Production in the Duplicating Unit of Department Y
D. Recommendation to Terminate Employee X's Services Because of Misconduct

13. Of the following, the MOST effective report writing style is usually characterized by 13.____
A. covering all the main ideas in the same paragraph
B. presenting each significant point in a new paragraph
C. placing the least important points before the most important points
D. giving all points equal emphasis throughout the report

14. Of the following, which factor is COMMON to all types of reports? 14.____
A. Presentation of information
B. Interpretation of findings
C. Chronological ordering of the information
D. Presentation of conclusions and recommendations

15. When writing a report, the one of the following which you should do FIRST is 15.____
A. set up a logical work schedule
B. determine your objectives in writing the report
C. select your statistical material
D. obtain the necessary data from the files

16. Good report writing utilizes, where possible, the use of table of contents, clear titles and sub-titles, well-labeled tables and figures, and good summaries in prominent places. 16.___
These features in a report are MOST helpful in
   A. saving the reader's time
   B. emphasizing objectivity
   C. providing a basic reference tool
   D. forming a basis for future action

17. The one of the following which BEST describes a periodic report is that it 17.___
   A. provides a record of accomplishments for a given time span and a comparison with similar time spans in the past
   B. covers the progress made in a project that has been postponed
   C. integrates, summarizes, and perhaps interprets published data on technical or scientific material
   D. describes a decision, advocates a policy or action, and presents facts in support of the writer's position

18. The PRIMARY purpose of including pictorial illustrations in a formal report is usually to 18.___
   A. amplify information which has been adequately treated verbally
   B. present details that are difficult to describe verbally
   C. provide the reader with a pleasant, momentary distraction
   D. present supplementary information incidental to the main ideas developed in the report

19. Of the following, which is usually the MOST important guideline in writing business letters? 19.___
A letter should be
   A. neat
   B. written in a formalized style
   C. written in clear language intelligible to the reader
   D. written in the past tense

20. Suppose you are asked to edit a policy statement. You note that personal pronouns like *you*, *we*, and *I* are used freely. 20.___
Which of the following statements BEST applies to this use of personal pronouns?
   A. It is proper usage because written business language should not be different from carefully spoken business language.
   B. It requires correction because it is ungrammatical.
   C. It is proper because it is clearer and has a warmer tone.
   D. It requires correction because policies should be expressed in an impersonal manner.

21. Good business letters are coherent. 21.____
To be coherent means to
A. keep only one unifying idea in the message
B. present the total message
C. use simple, direct words for the message
D. tie together the various ideas in the message

22. A functional forms file is a collection of forms which are grouped by 22.____
A. purpose B. department C. title D. subject

23. All of the following are reasons to consult a records retention schedule except one. 23.____
Which one is that?
To determine
A. whether something should be filed
B. how long something should stay in file
C. who should be assigned to filing
D. when something on file should be destroyed

24. A secretary is MOST likely to employ a form letter when 24.____
A. an answer is not required
B. the same information must be repeated from letter to letter
C. there is not enough information to write a detailed reply
D. varied correspondence must be sent out quickly

25. Of the following, the BASIC intent of naming a form is to provide the means to 25.____
A. code those factors recorded on each form
B. describe the use of the form
C. index each form
D. call attention to specific sections within each form

KEY (CORRECT ANSWERS)

1. B
2. D
3. D
4. C
5. B

6. C
7. A
8. D
9. D
10. A

11. D
12. C
13. B
14. A
15. B

16. A,C
17. A
18. B
19. C
20. D

21. D
22. A
23. C
24. B
25. B

TEST 2

DIRECTIONS: Each question or incomplete statement is followed by several suggested answers or completions. Select the one that BEST answers the question or completes the statement. *PRINT THE LETTER OF THE CORRECT ANSWER IN THE SPACE AT THE RIGHT.*

1. Assume that you are assigned the task of reducing the time and costs involved in completing a form that is frequently used in your agency. After analyzing the matter, you decide to reduce the writing requirements of the form through the use of ballot boxes and pre-printed data. 1.___
If exact copy-to-copy registration of this form is necessary, it is MOST advisable to
   A. vary the sizes of the ballot boxes
   B. stagger the ballot boxes
   C. place the ballot boxes as close together as possible
   D. have the ballot boxes follow the captions

2. To overcome problems that are involved in the use of cut-sheet and padded forms, specialty forms have been developed. Normally, these forms are commercially manufactured rather than produced in-plant. Before designing a form as a specialty form, however, you should be assured that certain factors are present. 2.___
Which one of the following factors deserve LEAST consideration?
   A. The form is to be used in quantities of 5,000 or more annually.
   B. The forms will be prepared on equipment using either a pinfeed device or pressure rollers for continuous feed-through.
   C. Two or more copies of the form set must be held together for further processing subsequent to the initial distribution of the form set.
   D. Copies of the form will be identical, and no items of data will be selectively eliminated from one or more copies of the form.

3. Although a well-planned form should require little explanation as to its completion, there are many occasions when the analyst will find it necessary to include instructions on the form to assure that the person completing it does so correctly. 3.___
With respect to such instructions, it is usually considered to be LEAST appropriate to place them
   A. in footnotes at the bottom of the form
   B. following the spaces to be completed
   C. directly under the form's title
   D. on the front of the form

4. One of the basic data-arrangement methods used in forms design is the *on-line* method. When this method is used, captions appear on the same line as the space provided for entry of the variable data. 4.____
This arrangement is NOT recommended because it
   A. forces the typist to make use of the typewriter's tab stops, thus increasing processing time
   B. wastes horizontal space since the caption appears on the writing line
   C. tends to make the variable data become more dominant than the captions
   D. increases the form's processing time by requiring the typist to continually roll the platen back and forth to expose the caption

5. Of the following, the BEST reason for using form letters in correspondence is that they are 5.____
   A. concise and businesslike
   B. impersonal in tone
   C. uniform in appearance
   D. economical for large mailings

6. Of the following, the MOST important reason to sort large volumes of documents before filing is that sorting 6.____
   A. decreases the need for cross-referencing
   B. eliminates the need to keep the files up to date
   C. prevents overcrowding of the file drawers
   D. saves time and energy in filing

7. To overcome the manual collation problem, forms are frequently padded. 7.____
Of the following statements which relate to this type of packaging, select the one that is MOST accurate.
   A. Typewritten forms which are prepared as padded forms are more efficient than all other packaging.
   B. Padded forms are best suited for handwritten forms.
   C. It is difficult for a printer to pad form copies of different colors.
   D. Registration problems increase when cut-sheet forms are padded.

8. Most forms are cut from a standard mill sheet of paper. This is the size on which forms dealers base their prices. Since an agency is paying for a full-size sheet of paper, it is the responsibility of the analyst to design forms so that as many as possible may be cut from the sheet without waste. 8.____
Of the following sizes, select the one that will cut from a standard mill sheet with the GREATEST waste and should, therefore, be avoided if possible.
   A. 4"x6"   B. 5"x8"   C. 9"x12"   D. $8\frac{1}{2}$"x14"

9. Assume that the work in your department involves the use of many technical terms. 9.____
In such a situation, when you are answering inquiries from the general public, it would usually be BEST to
A. use simple language and avoid the technical terms
B. use the technical terms whenever possible
C. use technical terms freely, but explain each term in parentheses
D. apologize if you are forced to use a technical term

10. You are answering a letter that was written on the letter-head of the ABC Company and signed by James H. Block, Treasurer. 10.____
What is usually considered to be the CORRECT salutation to use in your reply?
Dear
A. ABC Company:
B. Sirs:
C. Mr. Block:
D. Mr. Treasurer:

11. Assume that one of your duties is to handle routine letters of inquiry from the public. 11.____
The one of the following which is usually considered to be MOST desirable in replying to such a letter is a
A. detailed answer handwritten on the original letter of inquiry
B. phone call since you can cover details more easily over the phone than in a letter
C. short letter giving the specific information requested
D. long letter discussing all possible aspects of the questions raised

12. The CHIEF reason for dividing a letter into paragraphs is to 12.____
A. make the message clear to the reader by starting a new paragraph for each new topic
B. make a short letter occupy as much of the page as possible
C. keep the reader's attention by providing a pause from time to time
D. make the letter look neat and businesslike

13. An *Attention* line is used in correspondence to 13.____
A. indicate to the person receiving the correspondence that it contains an enclosure
B. direct correspondence addressed to an organization to a particular individual within the organization
C. greet the recipient of the correspondence
D. highlight the main concern of the correspondence

14. In deciding upon the advisability of recording certain information on a regular basis, the MOST important consideration is: 14.____
A. How much will it cost?
B. Is it necessary?
C. Is space available for keeping additional records?
D. Will it fit into the work pattern?

15. Instructions for filling out simple forms should USUALLY appear 15.___
   A. at the bottom of the form
   B. on a separate sheet of instructions
   C. on the reverse side of the form
   D. with the items to which they refer

16. Each new form should be given a number PRIMARILY because 16.___
   A. it provides a means of easy reference
   B. names are not sufficiently descriptive
   C. numbering forms is common government practice
   D. numbers are more suitable for automatic data processing

17. Of the following, the MOST important features of an effective business letter are 17.___
   A. introduction and conclusion
   B. punctuation and paragraphing
   C. simplicity and clarity
   D. style and organization

18. When recording receipt of purchases of equipment, the one of the following which is usually LEAST important is 18.___
   A. identification of the item
   B. name of the vendor
   C. quantity of the item
   D. weight of the item

19. In deciding which data should be collected for permanent records, the MOST important consideration is the 19.___
   A. amount of data available
   B. ease of processing the different types of data
   C. type of record-keeping system involved
   D. use to which such data may be put

20. In a certain filing system, documents are consecutively numbered as they are filed, a register is maintained of such consecutively numbered documents, and a record is kept of the number of each document removed from the files and its destination. 20.___
   This system will NOT help in
   A. finding the present whereabouts of a particular document
   B. proving the accuracy of the data recorded on a certain document
   C. indicating whether observed existing documents were ever filed
   D. locating a desired document without knowing what its contents are

21. The inside address in a business letter indicates to whom the letter is to be sent. 21.___
Of the following, the MOST important reason why a letter should contain the inside address is that the inside address
A. gives the letter a personal, friendly tone
B. simplifies the work of dictation and transcription
C. gives the letter a balanced appearance
D. identifies the addressee when the envelope containing the letter is discarded

22. The appearance of a business letter should make a favorable first impression on the person to whom the letter is sent. 22.___
In order to make such an impression, it is LEAST important that the
A. letter be centered on the page
B. margins be as even as possible
C. letter make a neat appearance
D. paragraphs be of the same length

23. A typed rough draft of a report should be double-spaced and should have wide margins PRIMARILY in order to 23.___
A. estimate the number of pages the report will contain
B. allow space for making corrections in the report
C. determine whether the report is well-organized
D. make the report easy to read

24. Suppose that you are assigned to make a number of original typewritten copies of a printed report. In doing this assignment, you type the first copy from the printed report and then type each subsequent copy from the last one you prepared. 24.___
You could be MOST certain that there were no errors made in the copies if you found no errors when comparing the
A. printed report with any one of the copies
B. first copy with the printed report
C. last copy with the printed report
D. first copy with the last copy

25. Before typing on more than one copy of a printed form, the one of the following which you should do FIRST is to 25.___
A. align the type so that the tails of the longer letters will rest on the lines printed on the form
B. check the alignment of the copies of the forms by holding them up to the light
C. insert the carbon paper into the typewriter and then insert the copies of the form
D. insert the copies of the form into the typewriter and then insert the carbon paper

KEY (CORRECT ANSWERS)

1. B
2. D
3. A
4. B
5. D

6. D
7. B
8. C
9. A
10. C

11. C
12. A
13. B
14. B
15. D

16. A
17. C
18. D
19. D
20. B

21. D
22. D
23. B
24. C
25. B

TEST 3

DIRECTIONS: Each question or incomplete statement is followed by several suggested answers or completions. Select the one that BEST answers the question or completes the statement. *PRINT THE LETTER OF THE CORRECT ANSWER IN THE SPACE AT THE RIGHT.*

1. The supervisor who makes a special point of using long words in preparing written reports is, in general, PROBABLY being 1.___
   A. *unwise* because a written report should be factual and accurate
   B. *unwise* because simplicity in a report is usually desirable
   C. *wise* because the written report will become a permanent record
   D. *wise* because with long words he can use the right emphasis in his report

2. Before you turn in a report you have written of an investigation that you made, you discover some additional information that you didn't know about before. 2.___
   Whether or not you rewrite your report to include this additional information should depend MAINLY on the
   A. amount of time left in which to submit the report
   B. effect this information will have on the conclusions of the report
   C. number of changes that you will have to make in your original report
   D. possibility of turning in a supplementary report later

3. When an applicant is approved for public assistance, the supervising clerk must fill in standard forms with certain information. 3.___
   The GREATEST advantage of using standard forms in this situation rather than having the supervising clerk write the report as he sees fit is that
   A. the report can be acted on quickly
   B. the report can be written without directions from a supervisor
   C. needed information is less likely to be left out of the report
   D. information that is written up this way is more likely to be verified

4. In some types of reports, visual aids add interest, meaning, and support. They also provide an essential means of effectively communicating the message of the report. 4.___
   Of the following, the selection of the suitable visual aids to use with a report is LEAST dependent on the
   A. nature and scope of the report
   B. way in which the aid is to be used

C. aids used in other reports
D. prospective readers of the report

5. A report is often revised several times before final preparation and distribution in an effort to make certain the report meets the needs of the situation for which it is designed. 5.____
Which of the following is the BEST way for the author to be sure that a report covers the areas he intended?
A. Obtain a co-worker's opinion
B. Compare it with a content checklist
C. Test it on a subordinate
D. Check his bibliography

6. Visual aids used in a report may be placed either in the text material or in the appendix. 6.____
Deciding where to put a chart, table, or any such aid should depend on the
A. title of the report
B. purpose of the visual aid
C. title of the visual aid
D. length of the report

7. In which of the following situations is an oral report PREFERABLE to a written report? 7.____
When a(n)
A. recommendation is being made for a future plan of action
B. department head requests immediate information
C. long-standing policy change is made
D. analysis of complicated statistical data is involved

8. All of the following rules will aid in producing clarity in report writing EXCEPT: 8.____
A. Give specific details or examples, if possible
B. Keep related words close together in each sentence
C. Present information in sequential order
D. Put several thoughts or ideas in each paragraph

9. When preparing a long report on a study prepared for your superior, the one of the following which should usually come FIRST in your report is a(n) 9.____
A. brief description of the working procedure followed in your study
B. review of the background conditions leading to the study
C. summary of your conclusions
D. outline of suggested procedures for implementing the report

10. The MAIN function of a research report is usually to 10.____
A. convince the reader of the adequacy of the research
B. report as expeditiously as possible what was done, why it was done, the results, and the conclusions
C. contribute to the body of scientific knowledge
D. substantiate an a priori conclusion by presenting a set of persuasive quantitative data

11. Words in a sentence must be arranged properly to make sure that the intended meaning of the sentence is clear. The sentence below that does NOT make sense because a clause has been separated from the word on which its meaning depends is: 11.___
   A. To be a good writer, clarity is necessary.
   B. To be a good writer, you must write clearly.
   C. You must write clearly to be a good writer.
   D. Clarity is necessary to good writing.

12. The use of a graph to show statistical data in a report is SUPERIOR to a table because it 12.___
   A. emphasizes approximations
   B. emphasizes facts and relationships more dramatically
   C. presents data more accurately
   D. is easily understood by the average reader

13. Of the following, the degree of formality required of a written report prepared by a labor relations specialist is MOST likely to depend on the 13.___
   A. subject matter of the report
   B. frequency of its occurrence
   C. amount of time available for its preparation
   D. audience for whom the report is intended

14. Of the following, a DISTINGUISHING characteristic of a written report intended for the head of your agency as compared to a report prepared for a lower-echelon staff member is that the report for the agency should usually include 14.___
   A. considerably more detail, especially statistical data
   B. the essential details in an abbreviated form
   C. all available source material
   D. an annotated bibliography

15. Assume that you are asked to write a lengthy report for use by the administrator of your agency, the subject of which is *The Impact of Proposed New Data Processing Operations on Line Personnel* in your agency. You decide that the most appropriate type of report for you to prepare is an analytical report, including recommendations. The MAIN reason for your decision is that 15.___
   A. the subject of the report is extremely complex
   B. large sums of money are involved
   C. the report is being prepared for the administrator
   D. you intend to include charts and graphs

16. Assume that you are preparing a report based on a survey dealing with the attitudes of employees in Division X regarding proposed new changes in compensating employees for working overtime. Three percent of the respondents to the survey voluntarily offer an unfavorable opinion on the method of assigning overtime work, a question not specifically asked of the employees. 16.___

On the basis of this information, the MOST appropriate and significant of the following comments for you to make in the report with regard to employees' attitudes on assigning overtime work is that

A. an insignificant percentage of employees dislike the method of assigning overtime work
B. three percent of the employees in Division X dislike the method of assigning overtime work
C. three percent of the sample selected for the survey voiced an unfavorable opinion on the method of assigning overtime work
D. some employees voluntarily voiced negative feelings about the method of assigning overtime work, making it impossible to determine the extent of this attitude

17. Four parts of a survey report are listed below, not necessarily in their proper order: 17.____

I. Body of report
II. Synopsis of report
III. Letter of transmittal
IV. Conclusions

Which one of the following represents the BEST sequence for inclusion of these parts in a report?

A. III, IV, I, II
B. II, I, III, IV
C. III, II, I, IV
D. I, III, IV, II

18. Of the following, the MOST important value of a good report is that it 18.____

A. reflects credit upon the person who submitted the report
B. provides good reference material
C. expedites official business
D. expresses the need for official action

19. The MOST important requirement in report writing is 19.____

A. promptness in turning in reports
B. length
C. grammatical construction
D. accuracy

20. You have discovered an error in your report submitted to the main office. 20.____
You should

A. wait until the error is discovered in the main office and then correct it
B. go directly to the supervisor in the main office after working hours and ask him unofficially to correct the answer
C. notify the main office immediately so that the error can be corrected if necessary
D. do nothing since it is possible that one error will have little effect on the total report

21. When you determine the methods of emphasis you will use in typing the titles, headings, and subheadings of a report, the one of the following which it is MOST important to keep in mind is that 21.___
   A. all headings of the same rank should be typed in the same way
   B. all headings should be typed in the single style which is most pleasing to the eye
   C. headings should not take up more than one-third of the page width
   D. only one method should be used for all headings, whatever their rank

22. Proper division of a letter into paragraphs requires that the writer of business letters should, as much as possible, be sure that 22.___
   A. each paragraph is short
   B. each paragraph develops discussion of just one topic
   C. each paragraph repeats the theme of the total message
   D. there are at least two paragraphs for every message

23. An editor is given a letter with this initial paragraph: 23.___
We have received your letter, which we read with interest, and we are happy to respond to your question. In fact, we talked with several people in our office to get ideas to send to you.
Which of the following is it MOST reasonable for the editor to conclude?
The paragraph is
   A. concise
   B. communicating something of value
   C. unnecessary
   D. coherent

24. In preparing a report that includes several tables, if not otherwise instructed, the typist should MOST properly include a list of tables 24.___
   A. in the introductory part of the report
   B. at the end of each chapter in the body of the report
   C. in the supplementary part of the report as an appendix
   D. in the supplementary part of the report as a part of the index

25. You have been asked to write a report on methods of hiring and training new employees. Your report is going to be about ten pages long. 25.___
For the convenience of your readers, a brief summary of your findings should
   A. appear at the beginning of your report
   B. be appended to the report as a postscript
   C. be circulated in a separate memo
   D. be inserted in tabular form in the middle of your report

26. A new student program is being set up for which certain new forms will be needed. You have been asked to design these forms. 26.____
Of the following, the FIRST step you should take in planning the forms is
   A. finding out the exact purpose for which each form will be used
   B. deciding what size of paper should be used for each form
   C. determining whether multiple copies will be needed for any of the forms
   D. setting up a new filing system to handle the new forms

27. Many government agencies require the approval by a central forms control unit of the design and reproduction of new office forms. 27.____
The one of the following results of this procedure that is a DISADVANTAGE is that requiring prior approval of a central forms control unit usually
   A. limits the distribution of forms to those offices with justifiable reasons for receiving them
   B. permits checking whether existing forms or modifications of them are in line with current agency needs
   C. encourages reliance on only the central office to set up all additional forms when needed
   D. provides for someone with a specialized knowledge of forms design to review and criticize new and revised forms

28. Suppose that you are assigned to prepare a form from which certain information will be posted in a ledger. 28.____
It would be MOST helpful to the person posting the information in the ledger if, in designing the form, you were to
   A. use the same color paper for both the form and the ledger
   B. make the form the same size as the pages of the ledger
   C. have the information on the form in the same order as that used in the ledger
   D. include in the form a box which is to be initialed when the data on the form have been posted in the ledger

29. In the effective design of office forms, the FIRST step to take is to 29.____
   A. decide what information should be included
   B. decide the purpose for which the form will be used
   C. identify the form by name and number
   D. identify the employees who will be using the form

30. Some designers of office forms prefer to locate the instructions on how to fill out the form at the bottom of it. 30.___
The MOST logical objection to placing such instructions at the bottom of the form is that
A. instructions at the bottom require an excess of space
B. all form instructions should be outlined with a separate paragraph
C. the form may be partly filled out before the instructions are seen
D. the bottom of the form should be reserved only for authorization and signature

KEY (CORRECT ANSWERS)

1. B	11. A	21. A
2. B	12. B	22. B
3. C	13. D	23. C
4. C	14. B	24. A
5. B	15. A	25. A
6. B	16. D	26. A
7. B	17. C	27. C
8. D	18. C	28. C
9. C	19. D	29. B
10. B	20. C	30. C

EXAMINATION SECTION

TEST 1

DIRECTIONS: Each question or incomplete statement is followed by several suggested answers or completions. Select the one that BEST answers the question or completes the statement.

Questions 1-17.

DIRECTIONS: In each of the following groups of sentences, there are three sentences which are correct and one which is incorrect because it contains an error in grammar, usage, diction, or punctuation. Indicate the letter of the INCORRECT sentence.

1. A. The business was organized under the name of Allen & Co. 1.___
   B. The price of admission was two dollars.
   C. The news was brought to Mr. Walters.
   D. There are less slips to be checked today than there were yesterday.

2. A. He only wants you to go with him; consequently I would be in the way. 2.___
   B. Whom do you think I saw on my way to lunch today?
   C. I am very much pleased with the work you are doing in my office.
   D. I think he is better than anyone else in his class.

3. A. I do not believe in his going so far away from home. 3.___
   B. She dresses exactly like her sister does.
   C. Neither Flora nor I is going to the movies tonight.
   D. The reason for my lateness is that the train was derailed.

4. A. I cannot understand its being on the bottom shelf because I remember putting it on the top shelf. 4.___
   B. If you do not agree with the statement above, please put a check next to it.
   C. We were both chosen to represent the association.
   D. The doctor assured us that she would not have to be operated.

5. A. Near the desk stand three chairs. 5.___
   B. How many crates of oranges were delivered?
   C. Where's your coat and hat?
   D. Either you or your mother is wrong.

6. A. She attacked the proposal with bitter words. 6.___
   B. Last year our team beat your team.
   C. The careless child spilled some milk on the table cloth.
   D. For three weeks last summer, Molly stood with her aunt.

7. A. Don't blame me for it. 7.____
   B. I have met but four.
   C. Loan me five dollars.
   D. May I leave early tonight?

8. A. It's time you knew how to divide by two numbers. 8.____
   B. Are you sure the bell has rung?
   C. Whose going to prepare the luncheon?
   D. Will it be all right if you are called at ten o'clock?

9. A. He had a wide knowledge of birds. 9.____
   B. New Orleans is further from Seattle than from Camden.
   C. Keats's poetry is characterized by rich imagery.
   D. He objected to several things - the cost, the gaudiness, and the congestion.

10. A. There was, in the first place, no indication that a crime had been committed. 10.____
    B. She is taller than any other member of the class.
    C. She decided to leave the book lay on the table.
    D. Haven't you any film in stock at this time?

11. A. Why do you still object to him coming with us to the party? 11.____
    B. If I were you, I should wait for them.
    C. If I were ten years older, I should like this kind of job.
    D. I shall go if you desire it.

12. A. Swimming in the pool, the water looked green. 12.____
    B. His speech is so precise as to seem affected.
    C. I would like to go overseas.
    D. We read each other's letters.

13. A. It must be here somewhere. 13.____
    B. The reason is that there is no bread.
    C. Of all other cities, New York is the largest.
    D. The sand was very warm at the beach.

14. A. If he were wealthy, he would build a hospital for the poor. 14.____
    B. I shall insist that he obey you.
    C. They saw that it was him.
    D. What kind of cactus is this one?

15. A. Because they had been trained for emergencies, the assult did not catch them by surprise. 15.____
    B. They divided the loot between the four of them in proportion to their efforts.
    C. The number of strikes is gradually diminishing.
    D. Between acts we went out to the lobby for a brief chat.

16. A. Through a ruse, the prisoners affected their escape from the concentration camp. 16.___
B. Constant esposure to danger has affected his mind.
C. Her affected airs served to alienate her from her friends.
D. Her vivacity was an affectation.

17. A. It is difficult to recollect that life was like before the war. 17.___
B. Will each of the pupils please hand their home work in?
C. There are fewer serious mistakes in this pamphlet than I had thought.
D. "Leave Her to Heaven" is the title of a novel by Ben Ames Williams.

Questions 18-25.

DIRECTIONS: Each of Questions 18 through 25 consists of three sentences lettered A, B, and C. In each of these questions, one of the sentences may contain an error in grammar, sentence structure, or punctuation, or all three sentences may be correct. If one of the sentences in a question contains an error in grammar, sentence structure, or punctuation, write in the space at the right the letter preceding the sentence which contains the error. If all three sentences are correct, write the letter D.

18. A. Mr. Smith appears to be less competent than I in performing these duties. 18.___
B. The supervisor spoke to the employee, who had made the error, but did not reprimand him.
C. When he found the book lying on the table, he immediately notified the owner.

19. A. Being locked in the desk, we were certain that the papers would not be taken. 19.___
B. It wasn't I who dictated the telegram; I believe it was Eleanor.
C. You should interview whoever comes to the office today.

20. A. The clerk was instructed to set the machine on the table before summoning the manager. 20.___
B. He said that he was not familiar with those kind of activities.
C. A box of pencils, in addition to erasers and blotters, was included in the shipment of supplies.

21. A. The supervisor remarked, "Assigning an employee to the proper type of work is not always easy." 21.___
B. The employer found that each of the applicants were qualified to perform the duties of the position.
C. Any competent student is permitted to take this course if he obtains the consent of the instructor.

22. A. The prize was awarded to the employee whom the judges believed to be most deserving. 22.___
    B. Since the instructor believes this book is the better of the two, he is recommending it for use in the school.
    C. It was obvious to the employees that the completion of the task by the scheduled date would require their working overtime.

23. A. These reports have been typed by employees who are trained by a capable supervisor. 23.___
    B. This employee is as old, if not older, than any other employee in the department.
    C. Running rapidly down the street, the messenger soon reached the office.

24. A. It is believed, that if these terms are accepted, the building can be constructed at a reasonable cost. 24.___
    B. The typists are seated in the large office; the stenographers, in the small office.
    C. Either the operators or the machines are at fault.

25. A. Mr. Jones, who is the head of the agency, will come today to discuss the plans for the new training program. 25.___
    B. The reason the report is not finished is that the supply of paper is exhausted.
    C. It is now obvious that neither of the two employees is able to handle this type of assignment.

KEY (CORRECT ANSWERS)

1. D
2. A
3. B
4. D
5. C

6. D
7. C
8. C
9. B
10. C

11. A
12. A
13. C
14. C
15. B

16. A
17. B
18. B
19. A
20. B

21. B
22. D
23. B
24. A
25. D

TEST 2

DIRECTIONS: In each of the following groups of sentences, one sentence is incorrect because it includes an error in grammar, usage, sentence structure, diction, capitalization, or punctuation. Indicate the INCORRECT sentence in each group.

1. A. We shall have to leave it to the jury to make a determination of the facts. 1.___
   B. His precision resulted in a nice discrimination between their relative merits.
   C. Green vegetables are healthy foods.
   D. We shall attempt to ascertain whether there has been any tampering with the lock.

2. A. Have you made any definitive plans which may be applied to budget preparation? 2.___
   B. We planned on taking a walking trip through the mountains.
   C. I would much rather he had called me after we had taken the trip.
   D. Do you believe that he has a predisposition toward that kind of response?

3. A. He carried out the orders with great dispatch but with little effect. 3.___
   B. The cook's overbearing manner overawed his employer.
   C. All of us shall partake of the benefits of exercise.
   D. Miss Smith made less errors than the other typists.

4. A. I believe that we are liable to have good weather tomorrow. 4.___
   B. From what I could see, I thought he acted like the others.
   C. Perpetual motion is an idea which is not unthinkable.
   D. Many of us taxpayers are displeased with the service.

5. A. She was incredulous when I told her the incredible tale. 5.___
   B. She was told that the symptoms would disappear within a week.
   C. If possible, I should like to sit in front of the very tall couple.
   D. Punish whomever disobeys our commands.

6. A. The men were trapped inside the cave for four days. 6.___
   B. The man seated in back of me was talking throughout the play.
   C. He told me that he doesn't know whether he will be able to visit us.
   D. Please bring me the pair of scissors from the table.

7. A. He was charged with having committed many larcenous acts. 7.____
B. Material wealth is certainly not something to be dismissed cavalierly.
C. He is one of those people who do everything promptly.
D. I hope to be able to retaliate for the assistance you have given me.

8. A. Have you noted the unusual phenomena to be seen in that portion of the heavens? 8.____
B. The data is as accurate as it is possible to make it.
C. The enormity of the crime was such that we could not comprehend it.
D. The collection of monies from some clients was long overdue.

9. A. What you are doing is not really different than what I had suggested. 9.____
B. The enormousness of the animal was enough to make her gasp.
C. The judge brought in a decision which aroused antagonism in the community.
D. I asked the monitor to take the papers to the principal.

10. A. He talks as if he were tired. 10.____
B. He amended his declaration to include additional income.
C. I know that he would have succeeded if he had tried.
D. Whom does Mrs. Jones think wrote the play?

11. A. The stone made a very angry bruise on his forearm. 11.____
B. He said to me: "I'm very made at you."
C. In all likelihood, we shall be unable to go to the fair.
D. He would have liked to go to the theatre with us.

12. A. The lawyers tried to settle the case out of court. 12.____
B. "Get out of my life!" she cried.
C. Walking down the road, the lake comes into view.
D. The loan which I received from the bank helped me to keep the business going.

13. A. I shall go with you providing that we return home early. 13.____
B. He has been providing us with excellent baked goods for many years.
C. It has been proved, to my satisfaction, to be correct.
D. Whether we go or not is for you to decide.

14. A. He does not seem able to present a logical and convincing argument. 14.____
B. Each of the goaltenders was trying to protect his respective cage.
C. He said, "I shall go there directly."
D. The reason he was late was on account of the delay in transportation.

15. A. Her mien revealed her abhorrence of his actions. 15.___
B. She used a great deal of rope so it would not come apart.
C. After he had lived among them, he found much to admire in their way of life.
D. He waited patiently for the fish to snatch at the bait.

16. A. As a result of constant exposure to the elements, he took sick and required medical attention. 16.___
B. Although the automobile is very old, we think it can still be used for a long trip.
C. He purchased all the supplies she requested with one exception.
D. He has repeated the story so frequently that I think he has begun to believe it.

17. A. It is the noise made by the crickets that you hear. 17.___
B. She told us that she would be at home on Sunday.
C. She said, "If I'm not there on time, don't wait on me."
D. Please try to maintain a cheerful disposition under any and all provocations.

18. A. Who is the tallest boy in the class? 18.___
B. Where shall I look to find a similar kind of stone?
C. The horse took the jumps with a great deal of ease.
D. He is as good, if not better than, any other jumper in the country.

19. A. Which of the two machines would be the most practical? 19.___
B. All of us are entitled to a reply if we are to determine whether you should remain as a member of the club.
C. Everyone who was listening got to his feet and applauded.
D. There was no indication from his actions that he knew he was wrong.

20. A. I beg leave to call upon you in case of emergency. 20.___
B. Do not deter me from carrying out the demands of my office.
C. Please see me irregardless of the time of day.
D. The intrepid captain shouted: "Into the fray!"

21. A. The teacher asked me whether she could lend the book from me. 21.___
B. You are not permitted to enter the public address system control room while an announcement is being made.
C. The mother announced loudly that she was going to the district office to find out whether we could refuse to accept her daughter.
D. It was my opinion that the salesman arrived at a most inopportune time to demonstrate his machine.

22. A. Perhaps we can eliminate any possibility of a misunderstanding by placing a special notice on the bulletin board. 22.___
B. The heavy snow storm caused a noticeable drop in student attendance.
C. We received two different relays, due to the fact that both the district office and the division office sent out separate notices.
D. Nevertheless, I urge you to prepare for the regular examination by taking the required courses.

23. A. By rotating the secretaries' tasks, we should be able to train the entire office staff in all duties. 23.___
B. The boy who had fallen told me that he felt alright, so that I didn't make out an accident report.
C. The flowers that the secretary had brought added a delightful touch of color to the office.
D. We shall have to work fast to complete the task before the deadline.

24. A. The telephone caller said that he was the boy's father, but his voice sounded immature. 24.___
B. There are very few situations which would require that we close the office and send the staff home early.
C. Although a packing slip accompanied the package, we had not received an invoice.
D. The parent claimed that the child had been in school, but the roll book indicated that the boy had been absent from school.

25. A. Needless to say, I could not grant the parent her request to see the teacher immediately since the teacher was teaching a class. 25.___
B. The small boy entered the office crying bitterly, and he refused to tell me the cause of his tears.
C. In describing her son, the mother told me that he was smaller than any boy in his class.
D. By holding teachers' checks until after lunch, you prevent many teachers from getting to the bank.

KEY (CORRECT ANSWERS)

1. C	6. B	11. B	16. A	21. A
2. B	7. D	12. C	17. C	22. C
3. D	8. B	13. A	18. D	23. B
4. A	9. A	14. D	19. A	24. D
5. D	10. D	15. B	20. C	25. C

TEST 3

DIRECTIONS: In each of the following groups of sentences, one of the four sentences contains one or more errors in grammar, sentence structure, English usage, or diction. Select the INCORRECT sentence in each case.

1. A. Protest as much as you like, I shall stick to my plan to the bitter end. 1.____
   B. It took two men to lift the refrigerator off of the truck.
   C. The dancer, with her company, her orchestra, and her manager, occupies the sixth floor.
   D. When one has worked with his hands he has really earned his keep.

2. A. The large number of crises among African governments indicate the difficulty of transition to independence. 2.____
   B. No sooner had the final bell rung than there was a mad scramble toward the door.
   C. Being as helpful as he could, the traffic policeman offered to send for a mechanic.
   D. They climbed higher that they might reach the little souvenir shop.

3. A. That angry retort of Father's came after long provocation. 3.____
   B. The frankness of the book presented a difficult problem as far as advertising it.
   C. It is the glory of Yale that she has many famous men to select from.
   D. The stars of the team were the following: Jones, center; Smith, end; and Harris, quarterback.

4. A. Next year we shall nominate whomever we please. 4.____
   B. The story is so well told that anything but the author's desired effect is impossible.
   C. My Oldsmobile has a Mercury motor, which makes it hard to shift gears.
   D. He is one of those people who believe in existentialist ideas.

5. A. After my previous experiences, I never expect to come this far again. 5.____
   B. I like my present work as preparing the way to my future occupation.
   C. To sit and smoke and think and dream was his idea of gratification.
   D. The snowball of knowledge sweeps relentlessly on, stamping additional rivets into the body of science.

6. A. The student racked his brains back and forth over the algebra problem. 6.____
B. It must be conceded that all the young men adapted themselves to the new regulations.
C. Formal talks were held between the three great powers in an effort to achieve disarmament.
D. The librarian felt bad about the damaged encyclopedia.

7. A. The policeman wanted to know if the driver of the car had a license. 7.____
B. Not many of these churches are less than thirty years old.
C. The athlete failed to break his unofficial world decathlon record by a narrow margin.
D. How different things appear in Washington than in London.

8. A. The agenda for the next meeting contains several highly important topics. 8.____
B. Stemming from the prelate's remarks was the inference that all was not well at the council.
C. Because the weight of majority opinion is so great is no reason why a dissenter should remain silent.
D. The general dominating the conference, there was no danger of chaos erupting.

9. A. Our company, which has thousands of employees, rates Jones one of its best men. 9.____
B. That wars go on may be considered somewhat of an inherited curse for posterity to bear.
C. Either Italian dressing or mayonnaise goes well with lettuce hearts.
D. Granted that he had the best intentions, his conduct was not above reproach.

10. A. There are two "i's" and two "e's" in privilege. 10.____
B. There is no use in Harry's brother saying anything about the situation.
C. I, not you, am to blame for the condition of this desk.
D. By relentless logic the group was lead to accept the statements of the eloquent stranger.

11. A. "It is one of those cars that go faster than 90 miles an hour," said the salesman. 11.____
B. Elementary school children are not the only ones who are tardy; it is also true of high school and college students.
C. My handsome (?) brother who is in college writes that he is "having a wonderful time"; he is, however, not doing too well scholastically.
D. A number of clerks were drinking coffee during the coffee break; the number of cups they drank was unbelievable.

12. A. He asked what had caused the accident. She replied that she did not know since she had not been present when it had occurred. 12.___
B. The new party headed by Prime Minister Wilson advocated government for the people, not by the people.
C. Five hundred yards of cloth are sufficient to do the work satisfactorily, and I plan to work continually until I finish.
D. She thought it would be all right for her to do the work in advance of the due date owing to the fact that none of the machines was being used by others.

13. A. "Oh! please stop that," he said; but when he looked up, they were nowhere to be seen. 13.___
B. The president of the company often told us workers of his experiences as a penniless, untrained beginner forty years ago.
C. The family showed its approval of the plan and decided to leave for Detroit, Michigan, on May 18th, or, if delay was unavoidable, on May 25th.
D. "I'm not buying," he said, "it's too expensive.

14. A. We expect in the next decade to more than hold our own in our race to the moon. 14.___
B. The teachers who have been considering the annual promotion plan today gave their report to the principal.
C. You are likely to find him sitting beside the brook in the park.
D. I shall have eaten by the time we go if my plans proceed according to schedule.

15. A. If you must know -- of course, however, this is a secret -- Billy just asked to borrow my car. 15.___
B. The job is over with, and neither my wife nor my children will ever persuade me to do it again.
C. You ought not to have said what you said, and I suggest that you apologize at once.
D. His friends had begun to understand how much he had done for the organization, but then they seemed to forget everything.

16. A. Unlike Bill and me, Ted looks really good in a stylish suit. 16.___
B. They have good teachers in our high schools; therefore, I plan to become a high school teacher.
C. Do you remember the name of the book? the author? the copyright date? the name of the publisher? of the editor?
D. The secretary and treasurer of the firm intends to hold a meeting with his president within a week or two.

17. A. Foreign films may be interesting, but I do not see them often. I usually prefer listening to music. 17.___
B. Court reporting has always fascinated me, but last spring I went to a lecture by a famous reporter. Then I made up my mind to be a reporter.
C. Haven't I asked you a hundred times to take the damaged typewriter to the repair shop which is in the store next to Jones's Candy Store?
D. Your typewriter should be kept absolutely clean and should be dusted as soon as you have completed your day's work.

18. A. The salesmen felt very pleased when they heard the manager say that their sales for the month of July, August, and September were much higher than that for the same months the previous year. 18.___
B. Four hundred dollars is too much to pay for these typewriters; I therefore suggest that you do some additional shopping before you make a final decision.
C. John said that he had swum around the lake three times and that he was now eligible for his swimming certificate.
D. I recently read "Trade Winds" in The Saturday Review.

19. A. Neither being sufficiently prepared for it, both brothers had to apply for additional training at the technical school. 19.___
B. The question was laid before them, and after weeks of argument it was still unsettled.
C. If all goes satisfactorily -- and why shouldn't it? -- we shall be in Europe before the middle of May
D. In his first year he was only an office boy, and in his fifth year he was president of the company.

20. A. Of all my friends he is the one on whom I can most surely depend. 20.___
B. We value the Constitution because of it's guarantees to freedom.
C. The audience was deeply stirred by the actor's performance.
D. Give the book to whoever comes into the room first.

21. A. Everything was in order: the paper rules, the pencils sharpened, the chairs placed. 21.___
B. Neither John nor Peter were able to attend the reception.
C. In April the streets which had been damaged by cold weather were repaired by the workmen.
D. You may lend my book to the pupil who you think will enjoy it most.

22. A. He fidgeted, like most children do, while the grown-ups were discussing the problem. 22.___
B. I won't go unless you go with me.
C. Sitting beside the charred ruins of his cabin, the frontiersman told us the story of the attack.
D. Certainly there can be no objection to the boys' working on a volunteer basis.

23. A. The congregation was dismissed. 23.___
B. The congregation were deeply moved by the sermon.
C. What kind of an automobile is that?
D. His explanation and mine agree.

24. A. There is no danger of him being elected. 24.___
B. There is no doubt of his election.
C. John and he are to be the speakers.
D. John and she are to be the speakers.

25. A. Them that honor me I will honor. 25.___
B. They that believe in me shall be rewarded.
C. Who did you see at the meeting?
D. Whom are you writing to?

KEY (CORRECT ANSWERS)

1. B
2. A
3. B
4. C
5. D

6. A
7. C
8. C
9. B
10. D

11. B
12. B
13. D
14. D
15. B

16. B
17. B
18. A
19. A
20. B

21. B
22. A
23. C
24. A
25. C

EXAMINATION SECTION

TEST 1

DIRECTIONS: In each of the following questions, only one of the four sentences conforms to standards of correct usage. The other three contain errors in grammar, diction, or punctuation. Select the choice in each question which BEST conforms to standards of correct usage. Consider a choice correct if it contains none of the errors mentioned above, even though there may be other ways of expressing the same thought. *PRINT THE LETTER OF THE CORRECT ANSWER IN THE SPACE AT THE RIGHT.*

1. A. Because he was ill was no excuse for his behavior. 1.____
   B. I insist that he see a lawyer before he goes to trial.
   C. He said "that he had not intended to go."
   D. He wasn't out of the office only three days.

2. A. He came to the station and pays a porter to carry his bags into the train. 2.____
   B. I should have liked to live in medieval times.
   C. My father was born in Linville. A little country town where everyone knows everyone else.
   D. The car, which is parked across the street, is disabled.

3. A. He asked the desk clerk for a clean, quiet, room. 3.____
   B. I expected James to be lonesome and that he would want to go home.
   C. I have stopped worrying because I have heard nothing further on the subject.
   D. If the board of directors controls the company, they may take actions which are disapproved by the stockholders.

4. A. Each of the players knew their place. 4.____
   B. He whom you saw on the stage is the son of an actor.
   C. Susan is the smartest of the twin sisters.
   D. Who ever thought of him winning both prizes?

5. A. An outstanding trait of early man was their reliance on omens. 5.____
   B. Because I had never been there before.
   C. Neither Mr. Jones nor Mr. Smith has completed his work.
   D. While eating my dinner, a dog came to the window.

6. A. A copy of the lease, in addition to the Rules and Regulations, are to be given to each tenant. 6.____
   B. The Rules and Regulations and a copy of the lease is being given to each tenant.
   C. A copy of the lease, in addition to the Rules and Regulations, is to be given to each tenant.
   D. A copy of the lease, in addition to the Rules and Regulations, are being given to each tenant.

7. A. Although we understood that for him music was a passion, we were disturbed by the fact that he was addicted to sing along with the soloists. 7.___
   B. Do you believe that Steven is liable to win a scholarship?
   C. Give the picture to whomever is a connoisseur of art.
   D. Whom do you believe to be the most efficient worker in the office?

8. A. Each adult who is sure they know all the answers will some day realize their mistake. 8.___
   B. Even the most hardhearted villain would have to feel bad about so horrible a tragedy.
   C. Neither being licensed teachers, both aspirants had to pass rigorous tests before being appointed.
   D. The principal reason why he wanted to be designated was because he had never before been to a convention.

9. A. Being that the weather was so inclement, the party has been postponed for at least a month. 9.___
   B. He is in New York City only three weeks and he has already seen all the thrilling sights in Manhattan and in the other four boroughs.
   C. If you will look it up in the official directory, which can be consulted in the library during specified hours, you will discover that the chairman and director are Mr. T. Henry Long.
   D. Working hard at college during the day and at the post office during the night, he appeared to his family to be indefatigable.

10. A. I would have been happy to oblige you if you only asked me to do it. 10.___
    B. The cold weather, as well as the unceasing wind and rain, have made us decide to spend the winter in Florida.
    C. The politician would have been more successful in winning office if he would have been less dogmatic.
    D. These trousers are expensive; however, they will wear well.

11. A. All except him wore formal attire at the reception for the ambassador. 11.___
    B. If that chair were to be blown off of the balcony, it might injure someone below.
    C. Not a passenger, who was in the crash, survived the impact.
    D. To borrow money off friends is the best way to lose them.

12. A. Approaching Manhattan on the ferry boat from Staten Island, an unforgettable sight of the skyscrapers is seen. 12.___
    B. Did you see the exhibit of modernistic paintings as yet?
    C. Gesticulating wildly and ranting in stentorian tones, the speaker was the sinecure of all eyes.
    D. The airplane with crew and passengers was lost somewhere in the Pacific Ocean.

13. A. If one has consistently had that kind of training, it is certainly too late to change your entire method of swimming long distances. 13.___
B. The captain would have been more impressed if you would have been more conscientious in evacuation drills.
C. The passangers on the stricken ship were all ready to abandon it at the signal.
D. The villainous shark lashed at the lifeboat with it's tail, trying to upset the rocking boat in order to partake of it's contents.

14. A. As one whose been certified as a professional engineer, I belive that the decision to build a bridge over that harbor is unsound. 14.___
B. Between you and me, this project ought to be completed long before winter arrives.
C. He fervently hoped that the men would be back at camp and to find them busy at their usual chores.
D. Much to his surprise, he discovered that the climate of Korea was like his home town.

15. A. An industrious executive is aided, not impeded, by having a hobby which gives him a fresh point of view on life and its problems. 15.___
B. Frequent absence during the calendar year will surely mitigate against the chances of promotion.
C. He was unable to go to the comittee meeting because he was very ill.
D. Mr. Brown expressed his disapproval so emphatically that his associates were embarassed.

16. A. At our next session, the office manager will have told you something about his duties and responsibilities. 16.___
B. In general, the book is absorbing and original and have no hesitation about recommending it.
C. The procedures followed by private industry in dealing with lateness and absence are different from ours.
D. We shall treat confidentially any information about Mr. Doe, to whom we understand you have sent reports to for many years.

17. A. I talked to one official, whom I knew was fully impartial. 17.___
B. Everyone signed the petition but him.
C. He proved not only to be a good student but also a good athlete.
D. All are incorrect.

18. A. Every year a large amount of tenants are admitted to housing projects. 18.___
B. Henry Ford owned around a billion dollars in industrial equipment.
C. He was aggravated by the child's bead behavior.
D. All are incorrect.

19. A. Before he was committed to the asylum he suffered from the illusion that he was Napoleon. 19.___
B. Besides stocks, there were also bonds in the safe.
C. We bet the other team easily.
D. All are incorrect.

20. A. Bring this report to your supervisor immediately. 20.___
B. He set the chair down near the table.
C. The capitol of New York is Albany.
D. All are incorrect.

21. A. He was chosen to arbitrate the dispute because everyone knew he would be disinterested. 21.___
B. It is advisable to obtain the best council before making an important decision.
C. Less college students are interested in teaching than ever before.
D. All are incorrect.

22. A. She, hearing a signal, the source lamp flashed. 22.___
B. While hearing a signal, the source lamp flashed.
C. In hearing a signal, the source lamp flashed.
D. As she heard a signal, the source lamp flashed.

23. A. Every one of the time records have been initialed in the designated spaces. 23.___
B. All of the time records has been initialed in the designated spaces.
C. Each one of the time records was initialed in the designated spaces.
D. The time records all been initialed in the designated spaces.

24. A. If there is no one else to answer the phone, you will have to answer it. 24.___
B. You will have to answer it yourself if no one else answers the phone.
C. If no one else is not around to pick up the phone, you will have to do it.
D. You will have to answer the phone when nobodys here to do it.

25. A. Dr. Barnes not in his office. What could I do for you? 25.___
B. Dr. Barnes is not in his office. Is there something I can do for you?
C. Since Dr. Barnes is not in his office, might there be something I may do for you?
D. Is there any ways I can assist you since Dr. Barnes is not in his office?

26. A. She do not understand how the new console works. 26.___
B. The way the new console works, she doesn't understand.
C. She doesn't understand how the new console works.
D. The new console works, so that she doesn't understand.

27. A. Certain changes in family income must be reported as they occur. 27.____
B. When certain changes in family income occur, it must be reported.
C. Certain family income changes must be reported as they occur.
D. Certain changes in family income must be reported as they have been occuring.

28. A. Each tenant has to complete the application themselves. 28.____
B. Each of the tenants have to complete the application by himself.
C. Each of the tenants has to complete the application himself.
D. Each of the tenants has to complete the application by themselves.

29. A. Yours is the only building that the construction will effect. 29.____
B. Your's is the only building affected by the construction.
C. The construction will only effect your building.
D. Yours is the only building that will be affected by the construction.

30. A. There is four tests left. 30.____
B. The number of tests left are four.
C. There are four tests left.
D. Four of the tests remains.

31. A. Each of the applicants takes a test. 31.____
B. Each of the applicants take a test.
C. Each of the applicants take tests.
D. Each of the applicants have taken tests.

32. A. The applicant, not the examiners, are ready. 32.____
B. The applicants, not the examiner, is ready.
C. The applicants, not the examiner, are ready.
D. The applicant, not the examiner, are ready.

33. A. You will not progress except you practice. 33.____
B. You will not progress without you practicing.
C. You will not progress unless you practice.
D. You will not progress provided you do not practice.

34. A. Neither the director or the employees will be at the office tomorrow. 34.____
B. Neither the director nor the employees will be at the office tomorrow.
C. Neither the director, or the secretary nor the other employees will be at the office tomorrow.
D. Neither the director, the secretary or the other employees will be at the office tomorrow.

35. A. In my absence he and her will have to finish the assignment. 35.___
B. In my absence he and she will have to finish the assignment.
C. In my absence she and him, they will have to finish the assignment.
D. In my absence he and her both will have to finish the assignment.

KEY (CORRECT ANSWERS)

1. B	11. A	21. A	31. A
2. B	12. D	22. D	32. C
3. C	13. C	23. C	33. C
4. B	14. B	24. A	34. B
5. C	15. A	25. B	35. B
6. C	16. C	26. C	
7. D	17. B	27. A	
8. B	18. D	28. C	
9. D	19. B	29. D	
10. D	20. B	30. C	

TEST 2

DIRECTIONS: Each question or incomplete statement is followed by several suggested answers or completions. Select the one that BEST answers the question or completes the statement. *PRINT THE LETTER OF THE CORRECT ANSWER IN THE SPACE AT THE RIGHT.*

Questions 1-4.

DIRECTIONS: Questions 1 through 4 consist of three sentences each. For each question, select the sentence which contains NO error in grammar or usage.

1. A. Be sure that everybody brings his notes to the conference. 1.____
   B. He looked like he meant to hit the boy.
   C. Mr. Jones is one of the clients who was chosen to represent the district
   D. All are incorrect.

2. A. He is taller than I. 2.____
   B. I'll have nothing to do with these kind of people.
   C. The reason why he will not buy the house is because it is too expensive.
   D. All are incorrect.

3. A. Aren't I eligible for this apartment. 3.____
   B. Have you seen him anywheres?
   C. He should of come earlier.
   D. All are incorrect.

4. A. He graduated college in 1982. 4.____
   B. He hadn't but one more line to write.
   C. Who do you think is the author of this report?
   D. All are incorrect.

Questions 5-35.

DIRECTIONS: In each of the following questions, only one of the four sentences conforms to standards of correct usage. The other three contain errors in grammar, diction, or punctuation. Select the choice in each question which BEST conforms to standards of correct usage. Consider a choice correct if it contains none of the errors mentioned above, even though there may be other ways of expressing the same thought.

5. A. It is obvious that no one wants to be a kill-joy if they can help it. 5.____
   B. It is not always possible, and perhaps it never is possible, to judge a person's character by just looking at him.

C. When Yogi Berra of the New York Yankees hit an immortal grand-slam home run, everybody in the huge stadium including Pittsburgh fans, rose to his feet.
D. Every one of us students must pay tuition today.

6. A. The physician told the young mother that if the baby is not able to digest its milk, it should be boiled. 6.___
B. There is no doubt whatsoever that he felt deeply hurt because John Smith had betrayed the trust.
C. Having partaken of a most delicious repast prepared by Tessie Breen, the hostess, the horses were driven home immediately thereafter.
D. The attorney asked my wife and myself several questions.

7. A. Despite all denials, there is no doubt in my mind that if my father were here you wouldn't talk like that. 7.___
B. At this time everyone must deprecate the demogogic attack made by one of our Senators on one of our most revered statesmen.
C. In the first game of a crucial two-game series, Ted Williams, got two singles, both of them driving in a run.
D. Our visitor brought good news to John and I.

8. A. If he would have told me, I should have been glad to help him in his dire financial emergency. 8.___
B. Newspaper men have often asserted that diplomats or so-called official spokesmen sometimes employ equivocation in attempts to deceive.
C. I think someones coming to collect money for the Red Cross.
D. In a masterly summation, the young attorney expressed his belief that the facts clearly militate aginst this opinion.

9. A. We have seen most all the exhibits. 9.___
B. Without in the least underestimating your advice, in my opinion the situation has grown immeasurably worse in the past few days.
C. I wrote to the box office treasurer of the hit show that a pair of orchestra seats would be preferable.
D. As the grim story of Pearl Harbor was broadcast on that fateful December 7, it was the general opinion that war was inevitable.

10. A. Without a moment's hesitation, Casey Stengel said that Larry Berra works harder than any player on the team. 10.___
B. There is ample evidence to indicate that many animals can run faster than any human being.
C. No one saw the accident but I.
D. Example of courage is the heroic defense put up by the paratroopers against overwhelming odds.

11. A. If you prefer these kind, Mrs. Grey, we shall be more than willing to let you have them reasonably. 11.____
B. If you like these here, Mrs. Grey, we shall be more than willing to let you have them reasonably.
C. If you like these, Mrs. Grey, we shall be more than willing to let you have them.
D. Who shall we appoint?

12. A. The number of errors are greater in speech than in writing. 12.____
B. The doctor rather than the nurse was to blame for his being neglected.
C. Because the demand for these books have been so great, we reduced the price.
D. John Galsworthy, the English novelist, could not have survived a serious illness; had it not been for loving care.

13. A. Our activities this year have seldom ever been as interesting as they have been this month. 13.____
B. Our activities this month have been more interesting, or at least as interesting as those of any month this year.
C. Our activities this month has been more interesting than those of any other month this year.
D. Neither Jean nor her sister was at home.

14. A. George B. Shaw's view of common morality, as well as his wit sparkling with a dash of perverse humor here and there, have led critics to term him "The Incurable Rebel." 14.____
B. The President's program was not always received with the wholehearted endorsement of his own party, which is why the party faces difficulty in drawing up a platform for the coming election.
C. The reason why they wanted to travel was because they had never been away from home.
D. Facing a barrage of cameras, the visiting celebrity found it extremely difficult to express his opinions clearly.

15. A. When we calmed down, we all agreed that our anger had been kind of unnecessary and had not helped the situation. 15.____
B. Without him going into all the details, he made us realize the horror of the accident.
C. Like one girl, for example, who applied for two positions.
D. Do not think that you have to be so talented as he is in order to play in the school orchestra.

16. A. He looked very peculiarly to me. 16.____
B. He certainly looked at me peculiar.
C. Due to the train's being late, we had to wait an hour.
D. The reason for the poor attendance is that it is raining.

17. A. About one out of four own an automobile. 17.___
B. The collapse of the old Mitchell Bridge was caused by defective construction in the central pier.
C. Brooks Atkinson was well acquainted with the best literature, thus helping him to become an able critic.
D. He has to stand still until the relief man comes up, thus giving him no chance to move about and keep warm.

18. A. He is sensitive to confusion and withdraws from people whom he feels are too noisy. 18.___
B. Do you know whether the data is statistically correct?
C. Neither the mayor or the aldermen are to blame:
D. Of those who were graduated from high school, a goodly percentage went to college.

19. A. Acting on orders, the offices were searched by a designated committee. 19.___
B. The answer probably is nothing.
C. I thought it to be all right to excuse them from class.
D. I think that he is as successful a singer, if not more successful, than Mary.

20. A. $120,000 is really very little to pay for such a well-built house. 20.___
B. The creatures looked like they had come from outer space.
C. It was her, he knew!
D. Nobody but me knows what to do.

21. A. Mrs. Smith looked good in her new suit. 21.___
B. New York may be compared with Chicago.
C. I will not go to the meeting except you go with me.
D. I agree with this editorial.

22. A. My opinions are different from his. 22.___
B. There will be less students in class now.
C. Helen was real glad to find her watch.
D. It had been pushed off of her dresser.

23. A. Almost everone, who has been to California, returns with glowing reports. 23.___
B. George Washington, John Adams, and Thomas Jefferson, were our first presidents.
C. Mr. Walters, whom we met at the bank yesterday, is the man, who gave me my first job.
D. One should study his lessons as carefully as he can.

24. A. We had such a good time yesterday. 24.___
B. When the bell rang, the boys and girls went in the schoolhouse.
C. John had the worst headache when he got up this morning.
D. Today's assignment is somewhat longer than yesterday's.

25. A. Neither the mayor nor the city clerk are willing to talk. 25.___
B. Neither the mayor nor the city clerk is willing to talk.
C. Neither the mayor or the city clerk are willing to talk.
D. Neither the mayor or the city clerk is willing to talk.

26. A. Being that he is that kind of boy, cooperation cannot be expected. 26.___
B. He interviewed people who he thought had something to say.
C. Stop whomever enters the building regardless of rank or office held.
D. Passing through the countryside, the scenery pleased us.

27. A. The childrens' shoes were in their closet. 27.___
B. The children's shoes were in their closet.
C. The childs' shoes were in their closet.
D. The childs' shoes were in his closet.

28. A. An agreement was reached between the defendant, the plaintiff, the plaintiff's attorney and the insurance company as to the amount of the settlement. 28.___
B. Everybody was asked to give their versions of the accident.
C. The consensus of opinion was that the evidence was inconclusive.
D. The witness stated that if he was rich, he wouldn't have had to loan the money.

29. A. Before beginning the investigation, all the materials relating to the case were carefully assembled. 29.___
B. The reason for his inability to keep the appointment is because of his injury in the accident.
C. This here evidence tends to support the claim of the defendant.
D. We interviewed all the witnesses who, according to the driver, were still in town.

30. A. Each claimant was allowed the full amount of their medical expenses. 30.___
B. Either of the three witnesses is available.
C. Every one of the witnesses was asked to tell his story.
D. Neither of the witnesses are right.

31. A. The commissioner, as well as his deputy and various bureau heads, were present. 31.___
B. A new organization of employers and employees have been formed.
C. One or the other of these men have been selected.
D. The number of pages in the book is enough to discourage a reader.

32. A. Between you and me, I think he is the better man. 32.___
B. He was believed to be me.
C. Is it us that you wish to see?
D. The winners are him and her.

33. A. Beside the statement to the police, the witness spoke to no one. 33.___
B. He made no statement other than to the police and I.
C. He made no statement to any one else, aside from the police.
D. The witness spoke to no one but me.

34. A. The claimant has no one to blame but himself. 34.___
B. The boss sent us, he and I, to deliver the packages.
C. The lights come from mine and not his car.
D. There was room on the stairs for him and myself.

35. A. Admission to this clinic is limited to patients' inability to pay for medical care. 35.___
B. Patients who can pay little or nothing for medical care are treated in this clinic.
C. The patient's ability to pay for medical care is the determining factor in his admissibility to this clinic.
D. This clinic is for the patient's that cannot afford to pay or that can pay a little for medical care.

KEY (CORRECT ANSWERS)

1. A	11. C	21. A	31. D
2. A	12. B	22. A	32. A
3. D	13. D	23. D	33. D
4. C	14. D	24. D	34. A
5. D	15. D	25. B	35. B
6. D	16. D	26. B	
7. B	17. B	27. B	
8. B	18. D	28. C	
9. D	19. B	29. D	
10. B	20. D	30. C	

English Expression

CHOICE OF EXPRESSION

COMMENTARY

One special form of the English Expression multiple-choice question in current use requires the candidate to select from among five (5) versions of a particular part of a sentence (or of an entire sentence), the one version that expresses the idea of the sentence most clearly, effectively, and accurately. Thus, the candidate is required not only to recognize errors, but also to choose the best way of phrasing a particular part of the sentence.

This is a test of choice of expression, which assays the candidate's ability to express himself correctly and effectively, including his sensitivity to the subleties and nuances of the language.

SAMPLE QUESTIONS

DIRECTIONS: In each of the following sentences some part of the sentence or the entire sentence is underlined. The underlined part presents a problem in the appropriate use of language. Beneath each sentence you will find five ways of writing the underlined part. The first of these indicates no change (that is, it repeats the original), but the other four are all different. If you think the original sentence is better than any of the suggested changes, you should choose answer A; otherwise you should mark one of the other choices. Select the best answer and blacken the corresponding space on the answer sheet.

This is a test of correctness and effectiveness of expression. In choosing answers, follow the requirements of standard written English; that is, pay attention to acceptable usage in grammar, diction (choice of words), sentence construction, and punctuation. Choose the answer that produces the most effective sentence -- clear and exact, without awkwardness or ambiguity. Do not make a choice that changes the meaning of the original sentence.

SAMPLE QUESTION 1

Although these states now trade actively with the West, and although they are willing to exchange technological information, their arts and thought, and social structure remains substantially similar to what it has always been.

A. remains substantially similar to what it has always been
B. remain substantially unchanged
C. remains substantially unchanged
D. remain substantially similar to what they have always been
E. remain substantially without being changed

The purpose of questions of this type is to determine the candidate's ability to select the clearest and most effective means of expressing what the statement attempts to say. In this example, the phrasing in the statement, which is repeated in **A**, presents a problem of agreement between a subject and its verb (their arts and thought and social structure and remains), a problem of agreement between a pronoun and its antecedent (their arts and thought and social structure and it), and a problem of precise and concise phrasing (remains substantially similar to what it has always been for remains substantially unchanged). Each of the four remaining choices in some way corrects one or more of the faults in the sentence, but only one deals with all three problems satisfactorily. Although C presents a more careful and concise wording of the phrasing of the statement and, in the process, eliminates the problem of agreement between pronoun and antecedent, it fails to correct the problem of agreement between the subject and its verb. In D, the subject agrees with its verb and the pronoun agrees with its antecedent, but the phrasing is not so accurate as it should be. The same difficulty persists in E. Only in B are all the problems presented corrected satisfactorily.

The question is not difficult.

SAMPLE QUESTION 2

Her latest novel is the largest in scope, the most accomplished in technique, and it is more significant in theme than anything she has written.

A. it is more significant in theme than anything
B. it is most significant in theme of anything
C. more significant in theme than anything
D. the most significant in theme than anything
E. the most significant in theme of anything

This question is of greater difficulty than the preceding one.

The problem posed in the sentence and repeated in A, is essentially one of parallelism: Does the underlined portion of the sentence follow the pattern established by the first two elements of the series (the largest ... the most accomplished)? It does not, for it introduces a pronoun and verb (it is) that the second term of the series indicates should be omitted and a degree of comparison (more significant) that is not in keeping with the superlatives used earlier in the sentence. B uses the superlative degree of significant but retains the unnecessary it is; C removes the it is, but retains the faulty comparative form of the adjective. D corrects both errors in parallelism, but introduces an error in idiom (the most ...than). Only E corrects all the problems without introducing another fault.

SAMPLE QUESTION 3

Desiring to insure the continuity of their knowledge, magical lore is transmitted by the chiefs to their descendants.

A. magical lore is transmitted by the chiefs
B. transmission of magical lore is made by the chiefs
C. the chiefs' magical lore is transmitted
D. the chiefs transmit magical lore
E. the chiefs make transmission of magical lore

The CORRECT answer is D.

SAMPLE QUESTION 4

As Malcolm walks quickly and confident into the purser's office, the rest of the crew wondered whether he would be charged with the theft.

A. As Malcolm walks quickly and confident
B. As Malcolm was walking quick and confident
C. As Malcolm walked quickly and confident
D. As Malcolm walked quickly and confidently
E. As Malcolm walks quickly and confidently

The CORRECT answer is D.

SAMPLE QUESTION 5

The chairman, granted the power to assign any duties to whoever he wished, was still unable to prevent bickering.

A. granted the power to assign any duties to whoever he wished
B. granting the power to assign any duties to whoever he wished
C. being granted the power to assign any duties to whoever he wished
D. having been granted the power to assign any duties to whosoever he wished
E. granted the power to assign any duties to whomever he wished

The CORRECT answer is E.

SAMPLE QUESTION 6

Certainly, well-seasoned products are more expensive, but those kinds prove cheaper in the end.

A. but those kinds prove cheaper
B. but these kinds prove cheaper
C. but that kind proves cheaper
D. but those kind prove cheaper
E. but this kind proves cheaper

The CORRECT answer is A.

SAMPLE QUESTION 7

"We shall not," he shouted, "whatever the difficulties," "lose faith in the success of our plan!"

A. difficulties," "lose faith in the success of our plan!"
B. difficulties, "lose faith in the success of our plan"!
C. "difficulties, lose faith in the success of our plan!"
D. difficulties, lose faith in the success of our plan"!
E. difficulties, lose faith in the success of our plan!"

The CORRECT answer is E.

SAMPLE QUESTION 8

Climbing up the tree, the lush foliage obscured the chattering monkeys.

A. Climbing up the tree
B. Having climbed up the tree
C. Clambering up the tree
D. After we had climbed up the tree
E. As we climbed up the tree

The CORRECT answer is E.

EXAMINATION SECTION

DIRECTIONS FOR THIS SECTION:
See DIRECTIONS for Sample Questions on page 1. *PRINT THE LETTER OF THE CORRECT ANSWER IN THE SPACE AT THE RIGHT.*

TEST 1

1. At the opening of the story, Charles Gilbert has just come to make his home with his two unmarried aunts. 1. ...
   A. No change B. hadn't hardly come C. has just came
   D. had just come E. has hardly came
2. The sisters, who are no longer young, are use to living quiet lives. 2. ...
   A. No change B. are used to live
   C. are use'd to living D. are used to living
   E. are use to live
3. They willingly except the child. 3. ...
   A. No change B. willingly eccepted
   C. willingly accepted D. willingly acepted
   E. willingly accept
4. As the months pass, Charles' presence affects many changes in their household. 4. ...
   A. No change B. affect many changes
   C. effects many changes D. effect many changes
   E. affected many changes
5. These changes is not all together to their liking. 5. ...
   A. No change B. is not altogether
   C. are not all together D. are not altogether
   E. is not alltogether
6. In fact, they have some difficulty in adapting theirselves to these changes. 6. ...
   A. No change B. in adopting theirselves
   C. in adopting themselves D. in adapting theirselves
   E. in adapting themselves
7. That is the man whom I believe was the driver of the car. 7. ...
   A. No change B. who I believed C. whom I believed
   D. who to believe E. who I believe
8. John's climb to fame was more rapid than his brother's. 8. ...
   A. No change B. than his brother
   C. than that of his brother's D. than for his brother
   E. than the brother
9. We knew that he had formerly swam on an Olympic team. 9. ...
   A. No change B. has formerly swum
   C. did formerly swum D. had formerly swum
   E. has formerly swam
10. Not one of us loyal supporters ever get a pass to a game 10. ...
    A. No change B. ever did got a pass
    C. ever has get a pass D. ever had get a pass
    E. ever gets a pass
11. He was complemented on having done a fine job. 11. ...
    A. No change B. was compliminted
    C. was compleminted D. was complimented
    E. did get complimented

12. This play is different from the one we had seen last night. 12. ...
    A. No change B. have seen C. had saw
    D. have saw E. saw
13. A row of trees was planted in front of the house. 13. ...
    A. No change B. was to be planted C. were planted
    D. were to be planted E. are planted
14. The house looked its age in spite of our attempts to beautify it. 14. ...
    A. No change B. looks its age C. looked its' age
    D. looked it's age E. looked it age
15. I do not know what to council in this case. 15. ...
    A. No change B. where to council
    C. when to councel D. what to counsel
    E. what to counsil
16. She is more capable than any other girl in the office. 16. ...
    A. No change B. than any girl
    C. than any other girls D. than other girl
    E. than other girls
17. At the picnic the young children behaved very good. 17. ...
    A. No change B. behave very good
    C. behaved better D. behave very well
    E. behaved very well
18. I resolved to go irregardless of the consequences. 18. ...
    A. No change B. to depart irregardless of
    C. to go regarding of D. to go regardingly of
    E. to go regardless of
19. The new movie has a number of actors which have been famous on Broadway. 19. ...
    A. No change B. which had been famous
    C. who had been famous D. that are famous
    E. who have been famous
20. I am certain that these books are not our's. 20. ...
    A. No change B. have not been ours'
    C. have not been our's D. are not ours
    E. are not ours'
21. Each of your papers is filed for future reference. 21. ...
    A. No change
    B. Each of your papers are filed
    C. Each of your papers have been filed
    D. Each of your papers are to be filed
    E. Each of your paper is filed
22. I wish that he would take his work more serious. 22. ...
    A. No change
    B. he took his work more serious
    C. he will take his work more serious
    D. he shall take his work more seriously
    E. he would take his work more seriously
23. After the treasurer report had been read, the chairman called for the reports of the committees. 23. ...
    A. No change
    B. After the treasure's report had been read
    C. After the treasurers' report had been read
    D. After the treasurerer's report had been read
    E. After the treasurer's report had been read

24. Last night the stranger lead us down the mountain. 24. ...
A. No change
B. leaded us down the mountain
C. let us down the mountain
D. led us down the mountain
E. had led us down the mountain

25. It would not be safe for either you or I to travel in Viet Nam. 25. ...
A. No change
B. for either you or me
C. for either I or you
D. for either of you or I
E. for either of I or you

TEST 2

1. Both the body and the mind needs exercise. 1. ...
A. No change
B. have needs of exercise
C. is needful of exercise
D. needed exercise
E. need exercise

2. It's paw injured, the animal limped down the road. 2. ...
A. No change
B. It's paw injured
C. Its paw injured
D. Its' paw injured
E. Its paw injure

3. The butter tastes rancidly. 3. ...
A. No change
B. tastes rancid
C. tasted rancidly
D. taste rancidly
E. taste rancid

4. Who do you think has sent me a letter? 4. ...
A. No change
B. Whom do you think
C. Whome do you think
D. Who did you think
E. Whom can you think

5. If more nations would have fought against tyranny, the course of history would have been different. 5. ...
A. No change
B. would fight
C. could have fought
D. fought
E. had fought

6. Radio and television programs, along with other media of communication, helps us to appreciate the arts and to keep informed. 6. ...
A. No change
B. helps us to appreciate the arts and to be informed
C. helps us to be appreciative of the arts and to keep informed
D. helps us to be appreciative of the arts and to be informed
E. help us to appreciate the arts and to keep informed

7. Music, for example most always has listening and viewing audiences numbering in the hundreds of thousands. 7. ...
A. No change
B. for example, most always
C. for example,almost always
D. for example nearly always
E. for example,near always

8. When operas are performed on radio or television, they effect the listener. 8. ...
A. No change
B. they inflict the listener
C. these effect the listeners
D. they affects the listeners
E. they affect the listener

9. After hearing them the listener wants to buy recordings of the music. 9. ...
   A. No change
   B. After hearing them, the listener wants
   C. After hearing them, the listener want
   D. By hearing them the listener wants
   E. By hearing them, the listener wants

10. To we Americans the daily news program has become important. 10. ...
   A. No change
   B. To we the Americans
   C. To us Americans
   D. To us the Americans
   E. To we and us Americans

11. This has resulted from it's coverage of a days' events. 11. ...
   A. No change
   B. from its coverage of a days' events
   C. from it's coverage of a day's events
   D. from its' coverage of a day's events
   E. from its coverage of a day's events

12. In schools, teachers advice their students to listen to or to view certain programs. 12. ...
   A. No change
   B. teachers advise there students
   C. teachers advise their students
   D. the teacher advises their students
   E. teachers advise his students

13. In these ways we are preceding toward the goal of an educated and an informed public. 13. ...
   A. No change
   B. we are preeceding toward the goal
   C. we are preceeding toward the goal
   D. we are proceding toward the goal
   E. we are proceeding toward the goal

14. The cost of living is raising again. 14. ...
   A. No change
   B. are raising again
   C. is rising again
   D. are rising again
   E. is risen again

15. We did not realize that the boys' father had forbidden them to keep there puppy. 15. ...
   A. No change
   B. had forbade them to keep there puppy
   C. had forbade them to keep their puppy
   D. has forbidden them to keep their puppy
   E. had forbidden them to keep their puppy

16. Her willingness to help others' was her outstanding characteristic. 16. ...
   A. No change
   B. Her willingness to help other's,
   C. Her willingness to help others's
   D. Her willingness to help others
   E. Her willingness to help each other

17. Because he did not have an invitation, the girls objected to him going, 17. ...
   A. No change
   B. the girls object to him going
   C. the girls objected to him's going
   D. the girls objected to his going
   E. the girls object to his going

18. Weekly dances have become a popular accepted feature of the summer schedule. 18. ...
   A. No change
   B. have become a popular accepted feature
   C. have become a popular excepted feature
   D. have become a popularly excepted feature
   E. have become a popularly accepted feature
19. I couldn't hardly believe that he would desert our party. 19. ...
   A. No change
   B. would hardly believe
   C. didn't hardly believe
   D. should hardly believe
   E. could hardly believe
20. I found the place in the book more readily than she. 20. ...
   A. No change
   B. more readily than her
   C. more ready than she
   D. more quickly than her
   E. more ready than her
21. A good example of American outdoor activities are sports. 21. ...
   A. No change
   B. is sports
   C. are sport
   D. are sports events
   E. are to be found in sports
22. My point of view is much different from your's. 22. ...
   A. No change
   B. much different than your's
   C. much different than yours
   D. much different from yours
   E. much different than yours'
23. The cook was suppose to use two spoonfuls of dressing for each serving. 23. ...
   A. No change
   B. was supposed to use two spoonsful
   C. was suppose to use two spoonsful
   D. was supposed to use two spoonsfuls
   E. was supposed to use two spoonfuls
24. If anyone has any doubt about the values of the tour, refer him to me. 24. ...
   A. No change
   B. refer him to I
   C. refer me to he
   D. refer them to me
   E. refer he to I
25. We expect that the affects of the trip will be beneficial. 25. ...
   A. No change
   B. the effects of the trip will be beneficial
   C. the effects of the trip should be beneficial
   D. the affects of the trip would be beneficial
   E. the effects of the trip will be benificial

TEST 3

1. That, my friend is not the proper attitude. 1. ...
   A. No change
   B. That my friend
   C. That my friend,
   D. That -- my friend
   E. That, my friend,
2. The girl refused to admit that the not was her's. 2. ...
   A. No change
   B. that the note were her's
   C. that the note was hers'
   D. that the note was hers
   E. that the note might be hers

3. There were fewer candidates that we had been lead to expect. 3. ...
   A. No change
   B. was fewer candidates than we had been lead
   C. were fewer candidates than we had been lead
   D. was fewer candidates than we had been led
   E. were fewer candidates than we had been led
4. When I first saw the car, its steering wheel was broke. 4. ...
   A. No change
   B. its' steering wheel was broken
   C. it's steering wheel had been broken
   D. its steering wheel were broken
   E. its steering wheel was broken
5. I find that the essential spirit for we beginners is missing. 5. ...
   A. No change
   B. we who begin are missing
   C. us beginners are missing
   D. us beginners is missing
   E. we beginners are missing
6. I believe that you had ought to study harder. 6. ...
   A. No change
   B. you should have ought
   C. you had better
   D. you ought to have
   E. you ought
7. This is Tom, whom I am sure, will be glad to help you. 7. ...
   A. No change
   B. Tom whom, I am sure,
   C. Tom, whom I am sure,
   D. Tom who I am sure,
   E. Tom, who, I am sure,
8. His father or his mother has read to him every night since he was very small. 8. ...
   A. No change
   B. did read to him
   C. have been reading to him
   D. had read to him
   E. have read to him
9. He become an authority on the theater and its great personalities. 9. ...
   A. No change
   B. becomed an authority
   C. become the authority
   D. became an authority
   E. becamed an authority
10. I know of no other person in the club who is more kind-hearted than her. 10. ...
    A. No change
    B. who are more kind-hearted than they
    C. who are more kind-hearted than them
    D. whom are more kind-hearted than she
    E. who is more kind-hearted than she
11. After Bill had ran the mile, he was breathless. 11. ...
    A. No change
    B. had runned the mile
    C. has ran the mile
    D. had ranned the mile
    E. had run the mile
12. Wilson has scarcely no equal as a pitcher. 12. ...
    A. No change
    B. has scarcely an equal
    C. has hardly no equal
    D. had scarcely no equal
    E. has scarcely any equals
13. It was the worse storm that the inhabitants of the island could remember. 13. ...
    A. No change
    B. were the worse storm
    C. was the worst storm
    D. was the worsest storm
    E. was the most worse storm

14. If only <u>we had began</u> before it was too late! 14. ...
    A. No change B. we had began
    C. we would have begun D. we had begun
    E. we had beginned
15. <u>Lets evaluate</u> our year's work. 15. ...
    A. No change B. Let us' evaluate
    C. Lets' evaluate D. Lets' us evaluate
    E. Let's evaluate
16. This is an organization <u>with which I wouldn't want to be associated with</u>. 16. ...
    A. No change
    B. with whom I wouldn't want to be associated with
    C. that I wouldn't want to be associated
    D. with which I would want not to be associated with
    E. with which I wouldn't want to be associated
17. The enemy fled in many directions, <u>leaving there weapons</u> on the field. 17. ...
    A. No change B. leaving its weapons
    C. letting their weapons D. leaving alone there weapons
    E. leaving their weapons
18. I hoped that John <u>could effect a compromise between</u> the approved forces. 18. ...
    A. No change
    B. could accept a compromise between
    C. could except a compromise between
    D. would have effected a compromise among
    E. could effect a compromise among
19. I was surprised to learn <u>that he has not always spoke English</u> fluently. 19. ...
    A. No change
    B. that he had not always spoke English
    C. that he did not always speak English
    D. that he has not always spoken English
    E. that he could not always speak English
20. The lawyer promised <u>to notify my father and I</u> of his plans for a new trial. 20. ...
    A. No change
    B. to notify I and my father
    C. to notify me and our father
    D. to notify my father and me
    E. to notify mine father and me
21. The most important feature of the series of tennis lessons <u>were the large amount</u> of strokes taught. 21. ...
    A. No change
    B. were the large number
    C. was the large amount
    D. was the largeness of the amount
    E. was the large number
22. That the prize proved to be beyond her reach <u>did not surprise him</u>. 22. ...
    A. No change
    B. has not surprised him
    C. had not ought to have surprised him
    D. should not surprise him
    E. would not have surprised him

23\. I am not all together in agreement with the author's point of view. 23. ...
A. No change B. all together of agreement
C. all together for agreement D. altogether with agreement
E. altogether in agreement

24\. Windstorms have recently established a record which meteorologists hope will not be equal for many years to come. 24. ...
A. No change B. will be equal
C. will not be equalized D. will be equaled
E. will not be equaled

25\. A large number of Shakespeare's soliloquies must be considered as representing thought, not speech. 25. ...
A. No change
B. as representative of speech, not thought
C. as represented by thought, not speech
D. as indicating thought, not speech
E. as representative of thought, more than speech

TEST 4

1\. A sight to inspire fear are wild animals on the lose. 1. ...
A. No change
B. are wild animals on the loose
C. is wild animals on the loose
D. is wild animals on the lose
E. are wild animals loose

2\. For many years, the settlers had been seeking to workship as they please. 2. ...
A. No change
B. had seeked to workship as they pleased
C. sought to workship as they please
D. sought to have worshiped as they pleased
E. had been seeking to worship as they pleased

3\. The girls stated that the dresses were their's. 3. ...
A. No change B. there's C. theirs D. theirs'
E. there own

4\. Please fellows don't drop the ball. 4. ...
A. No change B. Please, fellows C. Please fellows;
D. Please, fellows, E. Please! fellows

5\. Your sweater has laid on the floor for a week. 5. ...
A. No change B. has been laying
C. has been lying D. laid
E. has been lain

6\. I wonder whether you're sure that scheme of yours' will work. 6. ...
A. No change
B. your sure that scheme of your's
C. you're sure that scheme of yours
D. your sure that scheme of yours
E. you're sure that your scheme's

7\. Please let her and me do it. 7. ...
A. No change B. she and I C. she and me
D. her and I E. her and him

8\. I expected him to be angry and to scold her. 8. ...
   A. No change   B. and that he would scold
   C. and that he might scold   D. and that he should scold
   E. , scolding

9\. Knowing little about algebra, it was difficult to solve the equation. 9. ...
   A. No change
   B. the equation was difficult to solve
   C. the solution to the equation was difficult to find
   D. I found it difficult to solve the equation
   E. it being difficult to solve the equation

10\. He worked more diligent now that he had become vice president of the company. 10. ...
   A. No change   B. works more diligent
   C. works more diligently   D. began to work more diligent
   E. worked more diligently

11\. Flinging himself at the barricade he pounded on it furiously. 11. ...
   A. No change
   B. Flinging himself at the barricade: he
   C. Flinging himself at the barricade - he
   D. Flinging himself at the barricade; he
   E. Flinging himself at the barricade, he

12\. When he begun to give us advise, we stopped listening. 12. ...
   A. No change   B. began to give us advise
   C. begun to give us advice   D. began to give us advice
   E. begin to give us advice

13\. John was only one of the boys whom as you know was not eligible. 13. ...
   A. No change   B. who as you know were
   C. whom as you know were   D. who as you know was
   E. who as you know is

14\. Why was Jane and he permitted to go? 14. ...
   A. No change   B. was Jane and him
   C. were Jane and he   D. were Jane and him
   E. weren't Jane and he

15\. Take courage Tom: we all make mistakes. 15. ...
   A. No change   B. Take courage Tom - we
   C. Take courage, Tom; we   D. Take courage, Tom we
   E. Take courage! Tom: we

16\. Henderson, the president of the class and who is also captain of the team, will lead the rally. 16. ...
   A. No change
   B. since he is captain of the team
   C. captain of the team
   D. also being captain of the team
   E. who be also captain of the team

17\. Our car has always run good on that kind of gasoline. 17. ...
   A. No change   B. run well   C. ran good
   D. ran well   E. done good

18\. There was a serious difference of opinion among her and I. 18. ...
   A. No change   B. among she and I
   C. between her and I   D. between her and me
   E. among her and me

19. "This is most unusual," said Helen, "the mailman has never been this late before." 19. ...
    A. No change B. Helen, "The C. Helen - "The
    D. Helen; "The E. Helen." The
20. The three main characters in the story are Johnny Hobart a teenager, his mother a widow, and the local druggist. 20. ...
    A. No change
    B. teenager; his mother, a widow; and
    C. teenager; his mother a widow; and
    D. teenager, his mother, a widow and
    E. teenager, his mother,a widow; and
21. How much has food costs raised during the past year? 21. ...
    A. No change
    B. have food costs rose
    C. have food costs risen
    D. has food costs risen
    E. have food costs been raised
22. "Will you come too" she pleaded? 22. ...
    A. No change B. too,?"she pleaded.
    C. too?" she pleaded. D. too," she pleaded?
    E. too, she pleaded?"
23. If he would have drank more milk, his health would have been better. 23. ...
    A. No change B. would drink C. had drank
    D. had he drunk E. had drunk
24. Jack had no sooner laid down and fallen asleep when the alarm sounded. 24. ...
    A. No change
    B. no sooner lain down and fallen asleep than
    C. no sooner lay down and fell asleep when
    D. no sooner laid down and fell asleep than
    E. no sooner lain down than he fell asleep when
25. Jackson is one of the few Sophomores, who has ever made the varsity team. 25. ...
    A. No change
    B. one of the few Sophomores, who have
    C. one of the few sophomores, who has
    D. one of the few sophomores who have
    E. one of the few sophomores who has

TEST 5

1. The lieutenant had ridden almost a kilometer when the scattering shells begin landing uncomfortably close. 1. ...
    A. No change B. beginning to land C. began to land
    D. having begun to land E. begin to land
2. Having studied eight weeks, he now feels sufficiently prepared for the examination. 2. ...
    A. No change
    B. For eight weeks he studies so
    C. Due to eight weeks of study
    D. After eight weeks of studying
    E. Since he's been spending the last eight weeks in study

3. Coming from the Greek, and the word "democracy" means government by the people. 3. ...
   A. No change
   B. "Democracy," the word which comes from the Greek, means government by the people.
   C. Meaning government by the people, the word "democracy" comes from the Greek.
   D. Its meaning being government by the people in Greek, the word is "democracy."
   E. The word "democracy" comes from the Greek and means government by the people.
4. Moslem universities were one of the chief agencies in the development and spreading Arabic civilization. 4. ...
   A. No change B. in the development of
   C. to develop D. in developing
   E. for the developing of
5. The water of Bering Strait were closing to navigation by ice early in the fall. 5. ...
   A. No change B. has closed C. have closed
   D. had been closed E. closed
6. The man, since he grew up on the block, felt sentimental when returning to it. 6. ...
   A. No change B. having grown up
   C. growing up D. since he had grown up
   E. whose growth had been
7. Jack and Jill watched the canoe to take their parents out of sight round the bend of the creek. 7. ...
   A. No change
   B. The canoe, taking their parents out of sight, rounds the bend as Jack and Jill watch.
   C. Jack and Jill watched the canoe round the bend of the creek, taking their parents out of sight.
   D. The canoe rounded the bend of the creek as it took their parents out of sight, Jack and Jill watching.
   E. Jack and Jill watching,the canoe is rounding the bend of the creek to take their parents out of sight.
8. Chaucer's best-known work is THE CANTERBURY TALES, a collection of stories which he tells with a group of pilgrims as they travel to the town of Canterbury. 8. ...
   A. No change B. which he tells through C. who tell
   D. told by E. told through
9. The Estates-General, the old feudal assembly of France, had not met for one hundred and sevety-five years when it convened in 1789. 9. ...
   A. No change B. has not met C. has not been meeting
   D. had no meeting E. has no meeting
10. Just forty years ago, there had been fewer than one hundred symphony orchestras in the United States. 10. ...
    A. No change B. there had C. there were
    D. there was E. there existed
11. Mrs. Smith complained that her son's temper tantrums aggragravated her and caused her to have a headache. 11. ...
    A. No change B. gave her aggravation
    C. were aggravating to her D. aggravated her condition
    E. instigated

12. A girl like I would never be seen in a place like that. 12. ...
    A. No change B. as I C. as me D. like I am E. like me
13. Between you and me, my opinion is that this room is certainly nicer than the first one we saw. 13. ...
    A. No change B. between you and I C. among you and me
    D. betwixt you and I E. between we
14. It is important to know for what kind of a person you are working. 14. ...
    A. No change
    B. what kind of a person for whom you are working
    C. what kind of person you are working
    D. what kind of person you are working for
    E. what kind of a person you are working for
15. I had all ready finished the book before you came in. 15. ...
    A. No change B. already C. previously D. allready E. all
16. Ask not for who the bell tolls, it tolls for thee. 16. ...
    A. No change
    B. Ask not for whom the bell tolls, it tolls for thee.
    C. Ask not whom the bell tolls for; it tolls for thee.
    D. Ask not for whom the bell tolls; it tolls for thee.
    E. Ask not who the bell tolls for: It tolls for thee.
17. It is a far better thing I do, than ever I did before. 17. ...
    A. No change B. never I did C. I have ever did
    D. I have ever been done E. ever have I done
18. Ending a sentence with a preposition is something up with which I will not put. 18. ...
    A. No change
    B. Ending a sentence with a preposition is something with which I will not put up.
    C. To end a sentence with a preposition is that which I will not put up with.
    D. Ending a sentence with a preposition is something of which I will not put up.
    E. Something I will not put up with is ending a sentence with a preposition.
19. Everyone took off their hats and stand up to sing the national anthem. 19. ...
    A. No change
    B. took off their hats and stood up
    C. take off their hats and stand up
    D. took off his hat and stood up
    E. have taken off their hats and standing up
20. She promised me that if she had the opportunity she would have came irregardless of the weather. 20. ...
    A. No change
    B. She promised me that if she had the opportunity she would have come regardless of the weather.
    C. She assured me that had she had the opportunity she would have come regardless of the weather.
    D. She assured me that if she would have had the opportunity she would have come regardless of the weather.
    E. She promised me that if she had had the opportunity she would have came irregardless of the weather.

21. The man decided it would be advisable to marry a girl <u>somewhat younger than him</u>. 21. ...
    A. No change B. somehow younger than him
    C. some younger than him D. somewhat younger from him
    E. somewhat younger than he
22. Sitting near the campfire, the old man told <u>John and I about many exciting adventures he had had</u>. 22. ...
    A. No change
    B. John and me about many exciting adventures he had.
    C. John and I about much exciting adventure which he'd had.
    D. John and me about many exciting adventures he had had.
    E. John and me about many exciting adventures he has had.
23. <u>If you had stood at home and done your homework</u>, you would not have failed the course. 23. ...
    A. No change
    B. If you had stood at home and done you're homework,
    C. If you had staid at home and done your homework,
    D. Had you stayed at home and done your homework,
    E. Had you stood at home and done your homework,
24. The children didn't, as a rule, <u>do anything beyond</u> what they were told to do. 24. ...
    A. No change B. do hardly anything beyond
    C. do anything except D. do hardly anything except for
    E. do nothing beyond
25. <u>Either the girls or him is</u> right. 25. ...
    A. No change
    B. Either the girls or he is
    C. Either the girls or him are
    D. Either the girls or he are
    E. Either the girls nor he is

KEYS (CORRECT ANSWERS)

TEST 1

1.	A	11.	D
2.	D	12.	E
3.	E	13.	A
4.	C	14.	A
5.	D	15.	D
6.	E	16.	A
7.	E	17.	E
8.	A	18.	E
9.	D	19.	E
10.	E	20.	D

21. A
22. E
23. E
24. D
25. B

TEST 2

1.	E	11.	E
2.	C	12.	C
3.	B	13.	E
4.	A	14.	C
5.	E	15.	E
6.	E	16.	D
7.	C	17.	D
8.	E	18.	E
9.	B	19.	E
10.	C	20.	A

21. B
22. D
23. E
24. A
25. B

TEST 3

1.	E	11.	E
2.	D	12.	B
3.	E	13.	C
4.	E	14.	D
5.	D	15.	E
6.	E	16.	E
7.	E	17.	E
8.	A	18.	A
9.	D	19.	D
10.	E	20.	D

21. E
22. A
23. E
24. E
25. A

TEST 4

1.	C	11.	E
2.	E	12.	D
3.	C	13.	B
4.	D	14.	C
5.	C	15.	C
6.	C	16.	C
7.	A	17.	B
8.	A	18.	D
9.	D	19.	E
10.	E	20.	B

21. C
22. C
23. E
24. B
25. D

TEST 5

1.	C	11.	D
2.	A	12.	E
3.	E	13.	A
4.	D	14.	C
5.	D	15.	E
6.	B	16.	D
7.	C	17.	E
8.	D	18.	E
9.	A	19.	D
10.	C	20.	C

21. E
22. D
23. D
24. A
25. B

PREPARING WRITTEN MATERIAL

PARAGRAPH REARRANGEMENT

COMMENTARY

The sentences which follow are in scrambled order. You are to rearrange them in proper order and indicate the letter choice containing the correct answer at the space at the right.

Each group of sentences in this section is actually a paragraph presented in scrambled order. Each sentence in the group has a place in that paragraph; no sentence is to be left cut. You are to read each group of sentences and decide upon the best order in which to put the sentences so as to form as well-organized paragraph.

The questions in this section measure the ability to solve a problem when all the facts relevant to its solution are not given.

More specifically, certain positions of responsibility and authority require the employee to discover connections between events sometimes, apparently, unrelated. In order to do this, the employee will find it necessary to correctly infer that unspecified events have probably occurred or are likely to occur. This ability becomes especially important when action must be taken on incomplete information.

Accordingly, these questions require competitors to choose among several suggested alternatives, each of which presents a different sequential arrangement of the events. Competitors must choose the MOST logical of the suggested sequences.

In order to do so, they may be required to draw on general knowledge to infer missing concepts or events that are essential to sequencing the given events. Competitors should be careful to infer only what is essential to the sequence. The plausibility of the wrong alternatives will always require the inclusion of unlikely events or of additional chains of events which are NOT essential to sequencing the given events.

It's very important to remember that you are looking for the best of the four possible choices, and that the best choice of all may not even be one of the answers you're given to choose from.

There is no one right way to these problems. Many people have found it helpful to first write out the order of the sentences, as they would have arranged them, on their scrap paper before looking at the possible answers. If their optimum answer is there, this can save them some time. If it isn't, this method can still give insight into solving the problem. Others find it most helpful to just go through each of the possible choices, contrasting each as they go along. You should use whatever method feels comfortable, and works, for you.

While most of these types of questions are not that difficult, we've added a higher percentage of the difficult type, just to give you more practice. Usually there are only one or two questions on this section that contain such subtle distinctions that you're unable to answer confidently, and you then may find yourself stuck deciding between two possible choices, neither of which you're sure about.

EXAMINATION SECTION

TEST 1

DIRECTIONS: The sentences that follow are in scrambled order. You are to rearrange them in proper order and indicate the letter choice containing the correct answer. *PRINT THE LETTER OF THE CORRECT ANSWER IN THE SPACE AT THE RIGHT.*

1. Below are four statements labeled W.,X.,Y.,and Z. 1. ____
   W. He was a strict and fanatic drillmaster.
   X. The word is always used in a derogatory sense and generally shows resentment and anger on the part of the user.
   Y. It is from the name of this Frenchman that we derive our English word, martinet.
   Z. Jean Martinet was the Inspector-General of Infantry during the reign of King Louis XIV.

   The *PROPER* order in which these sentences should be placed in a paragraph is:
   A. X,Z,W,Y B. X,Z,Y,W C. Z,W,Y,X D. Z,Y,W,X

2. In the following paragraph, the sentences which are numbered, have been jumbled. 2. ____
   (1) Since then it has undergone changes.
   (2) It was incorporated in 1955 under the laws of the State of New York.
   (3) Its primary purpose, a cleaner city, has, however, remained the same.
   (4) The Citizens Committee works in cooperation with the Mayor's Inter-departmental Committee for a Clean City.

   The order in which these sentences should be arranged to form a well-organized paragraph is:
   A. 2,4,1,3 B. 3,4,1,2 C. 4,2,1,3 D. 4,3,2,1

Questions 3-5.

DIRECTIONS: The sentences listed below are part of a meaningful paragraph but they are not given in their proper order. You are to decide what would be the *best order* in which to put the sentences so as to form a well-organized paragraph. Each sentence has a place in the paragraph; there are no extra sentences. You are then to answer questions 3 to 5 inclusive on the basis of your rearrangements of these secrambled sentences into a properly organized paragraph.

In 1887 some insurance companies organized an Inspection Department to advise their clients on all phases of fire prevention and protection. Probably this has been due to the smaller annual fire losses in Great Britain than in the United States. It tests various fire prevention devices and appliances and determines manufacturing hazards and their safeguards. Fire research began earlier in the United States and is more advanced than in Great Britain. Later they established a laboratory specializing in electrical, mechanical, hydraulic, and chemical fields.

3. When the five sentences are arranged in proper order, the paragraph starts with the sentence which begins 3. ___
   A. "In 1887 ...." B. "Probably this ..." C. "It tests ..."
   D. "Fire research ..." E. "Later they ..."

4. In the last sentence listed above, "they" refers to 4. ___
   A. insurance companies B. the United States
   B. the United States and Great Britain
   C. the Inspection Department
   D. clients
   E. technicians

5. When the above paragraph is properly arranged, it ends with the words 5. ___
   A. "... and protection." B. "... the United States."
   C. "... their safeguards." D. "... in Great Britain."
   E. "... chemical fields."

KEY (CORRECT ANSWERS)

1. C
2. C
3. D
4. A
5. C

TEST 2

DIRECTIONS: In each of the questions numbered 1 through 5, several sentences are given. For each question, choose as your answer the group of numbers that represents the *most logical* order of these sentences if they were arranged in paragraph form. *PRINT THE LETTER OF THE CORRECT ANSWER IN THE SPACE AT THE RIGHT.*

1. 1. It is established when one shows that the landlord has prevented the tenant's enjoyment of his interest in the property leased. 1. ____
   2. Constructive eviction is the result of a breach of the covenant of quiet enjoyment implied in all leases.
   3. In some parts of the United States, it is not complete until the tenant vacates within a reasonable time.
   4. Generally, the acts must be of such serious and permanent character as to deny the tenant the enjoyment of his possessing rights.
   5. In this event, upon abandonment of the premises, the tenant's liability for that ceases.

   The CORRECT answer is:
   A. 2,1,4,3,5 B. 5,2,3,1,4 C. 4,3,1,2,5
   D. 1,3,5,4,2

2. 1. The powerlessness before private and public authorities that is the typical experience of the slum tenant is reminiscent of the situation of blue-collar workers all through the nineteenth century. 2. ____
   2. Similarly, in recent years, this chapter of history has been reopened by anti-poverty groups which have attempted to organize slum tenants to enable them to bargain collectively with their landlords about the conditions of their tenancies.
   3. It is familiar history that many of the workers remedied their condition by joining together and presenting their demands collectively.
   4. Like the workers, tenants are forced by the conditions of modern life into substantial dependence on these who possess great political and economic power.
   5. What's more, the very fact of dependence coupled with an absence of education and self-confidence makes them hesitant and unable to stand up for what they need from those in power.

   The CORRECT answer is:
   A. 5,4,1,2,3 B. 2,3,1,5,4 C. 3,1,5,4,2
   D. 1,4,5,3,2

3. 1. A railroad, for example, when not acting as a common carrier may contract away responsibility for its own negligence. 3. ___
2. As to a landlord, however, no decision has been found relating to the legal effect of a clause shifting the statutory duty of repair to the tenant.
3. The courts have not passed on the validity of clauses relieving the landlord of this duty and liability.
4. They have, however, upheld the validity of exculpatory clauses in other types of contracts.
5. Housing regulations impose a duty upon the landlord to maintain leased premises in safe condition.
6. As another example, a bailee may limit his liability except for gross negligence, willful acts, or fraud.

The CORRECT answer is:
A. 2,1,6,4,3,5 B. 1,3,4,5,6,2 C. 3,5,1,4,2,6
D. 5,3,4,1,6,2

4. 1. Since there are only samples in the building, retail or consumer sales are generally eschewed by mart occupants, and in some instances, rigid controls are maintained to limit entrance to the mart only to those persons engaged in retailing. 4. ___
2. Since World War I, in many larger cities, there has developed a new type of property, called the mart building.
3. It can therefore be used by wholesalers and jobbers for the display of sample merchandise.
4. This type of building is most frequently a multi-storied, finished interior property which is a cross between a retail arcade and a loft building.
5. This limitation enables the mart occupants to ship the orders from another location after the retailer or dealer makes his selection from the samples.

The CORRECT answer is:
A. 2,4,3,1,5 B. 4,3,5,1,2 C. 1,3,2,4,5
D. 1,4,2,3,5

5. 1. In general, staff-line friction reduces the distinctive contribution of staff personnel. 5. ___
2. The conflicts, however, introduce an uncontrolled element into the managerial system.
3. On the other hand, the natural resistance of the line to staff innovations probably usefully restrains over-eager efforts to apply untested procedures on a large scale.
4. Under such conditions, it is difficult to know when valuable ideas are being sacrificed.
5. The relatively weak position of staff, requiring accommodation to the line, tends to restrict their ability to engage in free, experimental innovation.

The CORRECT answer is:
A. 4,2,3,1,3 B. 1,5,3,2,4 C. 5,3,1,2,4
D. 2,1,4,5,3

KEY (CORRECT ANSWERS)

1. A
2. D
3. D
4. A
5. B

TEST 3

DIRECTIONS: Questions 1 through 4 consist of six sentences which can be arranged in a logical sequence. For each question, select the choice which places the numbered sentences in the *most logical* sequence. *PRINT THE LETTER OF THE CORRECT ANSWER IN THE SPACE AT THE RIGHT.*

1\. 1. The burden of proof as to each issue is determined before trial and remains upon the same party throughout the trial. 1. ___

2. The jury is at liberty to believe one witness' testimony as against a number of contradictory witnesses.
3. In a civil case, the party bearing the burden of proof is required to prove his contention by a fair preponderance of the evidence.
4. However, it must be noted that a fair preponderance of evidence does not necessarily mean a greater number of witnesses.
5. The burden of proof is the burden which rests upon one of the parties to an action to persuade the trier of the facts, generally the jury, that a proposition he asserts is true.
6. If the evidence is equally balanced, or if it leaves the jury in such doubt as to be unable to decide the controversy either way, judgment must be given against the party upon whom the burden of proof rests.

The CORRECT answer is:
A. 3,2,5,4,1,6 B. 1,2,6,5,3,4 C. 3,4,5,1,2,6
D. 5,1,3,6,4,2

2\. 1. If a parent is without assets and is unemployed, he cannot be convicted of the crime of non-support of a child. 2. ___

2. The term "sufficient ability" has been held to mean sufficient financial ability.
3. It does not matter if his unemployment is by choice or unavoidable circumstances.
4. If he fails to take any steps at all, he may be liable to prosecution for endangering the welfare of a child.
5. Under the penal law, a parent is responsible for the support of his minor child only if the parent is "of sufficient ability."
6. An indigent parent may meet his obligation by borrowing money or by seeking aid under the provisions of the Social Welfare Law.

The CORRECT answer is:
A. 6,1,5,3,2,4 B. 1,3,5,2,4,6 C. 5,2,1,3,6,4
D. 1,6,4,5,2,3

3. 1. Consider, for example, the case of a rabble rouser who urges a group of twenty people to go out and break the windows of a nearby factory. 3. ___
   2. Therefore, the law fills the indicated gap with the crime of "inciting to riot."
   3. A person is considered guilty of inciting to riot when he urges ten or more persons to engage in tumultuous and violent conduct of a kind likely to create public alarm.
   4. However, if he has not obtained the cooperation of at least four people, he cannot be charged with unlawful assembly.
   5. The charge of inciting to riot was added to the law to cover types of conduct which cannot be classified as either the crime of "riot" or the crime of "unlawful assembly."
   6. If he acquires the acquiescence of at least four of them, he is guilty of unlawful assembly even if the project does not materialize.

   The CORRECT answer is:
   A. 3,5,1,6,4,2 B. 5,1,4,6,2,3 C. 3,4,1,5,2,6
   D. 5,1,4,6,3,2

4. 1. If, however, the rebuttal evidence presents an issue of credibility, it is for the jury to determine whether the presumption has, in fact, been destroyed. 4. ___
   2. Once sufficient evidence to the contrary is introduced, the presumption disappears from the trial.
   3. The effect of a presumption is to place the burden upon the adversary to come forward with evidence to rebut the presumption.
   4. When a presumption is overcome and ceases to exist in the case, the fact or facts which gave rise to the presumption still remain.
   5. Whether a presumption has been overcome is ordinarily a question for the court.
   6. Such information may furnish a basis for a logical inference.

   The CORRECT answer is:
   A. 4,6,2,5,1,3 B. 3,2,5,1,4,6 C. 5,3,6,4,2,1
   D. 5,4,1,2,6,3

KEY (CORRECT ANSWERS)

1. D
2. C
3. A
4. B

EXAMINATION SECTION

DIRECTIONS: Each group of sentences in this section is actually a paragraph presented in scrambled order. Each sentence in the group has a place in that paragraph; no sentence is to be left out. You are to read each group of sentences so as to form a well-organized paragraph. Before trying to answer the questions which follow each group of sentences, jot down the correct order of the sentences. Then answer each of the questions by printing the letter of the correct answer in the space at the right. Remember that you will receive credit only for answers marked.

TEST 1

P. It is unfounded because, while the weak resent the power of the strong, they also respect it.
Q. The hesitancy stems from a concern for public opinion in other countries.
R. The United States has ordinarily been ill at ease in using its military power in support of its interests.
S. The concern is largely unfounded.
T. The roots of American hesitancy are deeply imbedded in the American mind.

1. Which sentence did you put last? 1.____
   A. P B. Q C. R D. S E. T

2. Which sentence did you put after Sentence R? 2.____
   A. P
   B. Q
   C. S
   D. T
   E. None of the above. Sentence R is last.

3. Which sentence did you put before Sentence S? 3.____
   A. P
   B. Q
   C. R
   D. T
   E. None of the above. Sentence S is first.

4. Which sentence did you put before Sentence R? 4.____
   A. P
   B. Q
   C. S
   D. T
   E. None of the above. Sentence R is last.

5. Which sentence did you put fourth? 5.____
   A. P B. Q C. R D. S E. T

TEST 2

P. Its lawlessness was virtually non-existent.
Q. The *Old West*, as portrayed in motion pictures, on television, and in books, is completely distorted.
R. It is obvious, therefore, that the *Old West* is falsely presented in mass media solely for commercial purposes.
S. Its heroes, too, were far from heroic.
T. Those who lived in the *Old West* in its final days, or talked to oldtimers, know the truth.

1. Which sentence did you put last? 1.____
   A. P B. Q C. R D. S E. T

2. Which sentence did you put after Sentence Q? 2.____
   A. P
   B. R
   C. S
   D. T
   E. None of the above. Sentence Q is last.

3. Which sentence did you put before Sentence S? 3.____
   A. P
   B. Q
   C. R
   D. T
   E. None of the above. Sentence S is first.

4. Which sentence did you put before Sentence Q? 4.____
   A. P
   B. R
   C. S
   D. T
   E. None of the above. Sentence Q is first.

5. Which sentence did you put after Sentence S? 5.____
   A. P
   B. Q
   C. R
   D. T
   E. None of the above. Sentence S is last.

TEST 3

P. One advertising executive became agitated recently when he suddenly realized that the floors of supermarkets were being unimaginatively used merely to walk on.
Q. Blank spaces, advertising men feel, cry out to be filled with merchandise-hustling messages.
R. He invented a slide projector which projects images on sheets of translucent plastic imbedded in supermarket floors.
S. At once, he got to work to correct this unforgiveable oversight.
T. As nature abhors a vacuum, so do advertising men decry blank spaces.

1. Which sentence did you put last? 1.____
   A. P
   B. Q
   C. S
   D. T
   E. None of the above. Sentence R is last.

2. Which sentence did you put third? 2.____
   A. P B. Q C. R D. S E. T

3. Which sentence did you put before Sentence T? 3.____
   A. P
   B. Q
   C. R
   D. T
   E. None of the above. Sentence T is first.

4. Which sentence did you put after Sentence P? 4.____
   A. Q
   B. R
   C. S
   D. T
   E. None of the above. Sentence P is last.

5. Which sentence did you put before Sentence Q? 5.____
   A. P
   B. R
   C. S
   D. T
   E. None of the above. Sentence Q is last.

TEST 4

P. It is estimated that Americans smoked almost a trillion cigarettes in 1966, while they smoked only several hundred million cigars and pipefuls of tobacco.
Q. Originally, they were considered exclusively a *ladies'* smoke.
R. Only in this century did cigarettes become popular in the United States.
S. Far more Americans smoke cigarettes today than smoke cigars and pipes combined.
T. This was not always the case, however.

1. Which sentence did you put first? 1.____
   A. P B. Q C. R D. S E. T

2. Which sentence did you put after Sentence Q? 2.____
   A. P
   B. R
   C. S
   D. T
   E. None of the above. Sentence Q is last.

3. Which sentence did you put before Sentence T? 3.____
   A. P
   B. Q
   C. R
   D. S
   E. None of the above. Sentence T is last.

4. Which sentence did you put after Sentence R? 4.____
   A. P
   B. Q
   C. S
   D. T
   E. None of the above. Sentence R is last.

5. Which sentence did you put before Sentence R? 5.____
   A. P
   B. Q
   C. S
   D. T
   E. None of the above. Sentence R is first.

TEST 5

P. A *megagram*, or a million *grams*, is, therefore, equal to 2,205 pounds.
Q. A *gram* is equivalent to 1/28.35 ounces.
R. The fundamental unit of mass in the metric system is the *gram*.
S. A *kilogram*, or a thousand *grams*, is equal to 2.205 pounds.
T. *Gram* is derived from the Late Greek, *gramma*, meaning *a small weight*.

1. Which sentence did you put after Sentence S? 1.____
   A. P
   B. Q
   C. R
   D. T
   E. None of the above. Sentence S is last.

2. Which sentence did you put before Sentence T? 2.____
   A. P
   B. Q
   C. R
   D. S
   E. None of the above. Sentence T is first.

3. Which sentence did you put after Sentence Q? 3.____
   A. P
   B. R
   C. S
   D. T
   E. None of the above. Sentence Q is last.

4. Which sentence did you put before Sentence R? 4.____
   A. P
   B. Q
   C. S
   D. T
   E. None of the above. Sentence R is first.

5. Which sentence did you put after Sentence T? 5.____
   A. P
   B. Q
   C. R
   D. S
   E. None of the above. Sentence T is last.

KEY (CORRECT ANSWERS)

TEST 1	TEST 2	TEST 3	TEST 4	TEST 5
1. A	1. C	1. E	1. D	1. A
2. D	2. D	2. A	2. E	2. C
3. B	3. A	3. E	3. A	3. C
4. E	4. E	4. C	4. B	4. E
5. D	5. C	5. D	5. D	5. B

BASIC FUNDAMENTALS OF WRITTEN COMMUNICATION

CONTENTS

BASIC FUNDAMENTALS OF WRITTEN COMMUNICATION

INSTRUCTIONAL OBJECTIVES

1. Ability to write legibly.
2. Ability to fill out forms and applications correctly.
3. Ability to take messages and notes accurately.
4. Ability to write letters effectively.
5. Ability to write directions and instructions clearly.
6. Ability to outline written and spoken information.
7. Ability to persuade or teach others through written communication.
8. Ability to write effective overviews and summaries.
9. Ability to make smooth transitions within written communications.
10. Ability to use language forms appropriate for the reader.
11. Ability to prepare effective informational reports.

CONTENT

INTRODUCTION

Public-service employees are required to prepare written communications for a variety of purposes. Written communication is a fundamental tool, not only for the public-service occupations, but throughout the world of work. Many public-service occupations require written communication with ordinary citizens of diverse backgrounds, so the trainee should develop the ability to write in simple, nontechnical language that the ordinary citizen will understand.

This unit is designed to develop the student's ability to communicate effectively in writing for a number of different purposes and in a number of different formats. Whatever the particular purpose or format, however, effective writing will require the writer:

- to have a clear idea of his purpose and his audience;
- to organize his thoughts and information in an orderly way;
- to express himself concisely, accurately, and concretely;
- to report relevant facts;
- to explain and summarize ideas clearly; and
- to evaluate the effectiveness of his communication.

1. BUSINESS WRITING

Several forms of written communication tend to recur frequently in most public-service agencies, including:

- letters
- forms
- memoranda
- minutes of meetings
- short reports
- telegrams and cables
- news releases
- and many others

The public-service employee should be familiar with the principles of writing in these forms, and should be able to apply them in preparing effective communications.

Letters

Every letter sent from a public-service agency should be considered an ambassador of goodwill. The impression it creates may mean the difference between favorable public attitudes or unfavorable ones. It may mean the difference between creating a friend or an enemy for the agency. Every public-service employee has a responsibility to serve the public effectively and to provide services in an efficient and courteous manner. The letters an agency sends out reflect its attitudes toward the public.

The impression a letter creates depends upon both its appearance and its tone. A letter which shows erasures and pen written corrections gives an impression that the sending agency is slovenly. Similarly, a rude or impersonal letter creates the impression that the agency is insensitive or unfeeling. In preparing letters, the employee should apply principles of style and tone which will serve to create the most favorable impression.

Select the Letter Type. The two most common types of business letters are letters of inquiry and letters of response - that is, "asking" letters and "answering" letters. Whichever type of letter the employee is asked to write, the following guidelines will simplify the task and help to achieve a style and tone which will create a favorable impression on the reader.

Select the Right Format. Several styles of letter format are in common use today, including:

- the indented format,
- the block format, and
- the semi-block format.

Modified forms of these are also in use in some offices. The student should become familiar with the formats preferred for usage in his office, and be able to use whichever form the employer requests.

Know the Letter Elements. Every letter includes certain basic elements, such as:

- the letterhead, which identifies the name and address of the sender.
- the date on which the letter was transmitted.
- the inside address, with the name, street, city, and state of the addressee.
- the salutation, greeting the addressee.
- the body, containing the message.
- the complimentary close, the "good-bye" of the business letter.
- the signature, handwritten by the sender.
- the typed signature, the typewritten name and title of the sender.

In addition, several other elements are occasionally found in business letters:

- the *attention line*, directing the letter to the attention of a particular individual or his representative.
- the *subject line*, informing the reader at a glance of the subject of the letter.

- the *enclosure notation*, noting items enclosed with the letter.
- the *carbon-copy notation*, listing other persons who receive copies of the letter.
- the *postscript*, an afterthought sometimes (but not normally) added following the last typed line of the letter.

Be Brief. Use only the words which help to say what is needed in a clear and straightforward manner. Do not repeat information already known to the reader, or contained elsewhere in the letter. Likewise, do not repeat information contained in the letter being answered. Rather than repeat the content of a previous letter, one can say something like, "Please refer to our letter dated March 5."

An employee can shorten his letters by using single words that serve the same function as longer phrases. Many commonly used phrases can be replaced by single words. For example,

Phrase	Single word
in order to	to
in reference to	about
in the amount of	for, of
in a number of cases	some
in view of	because
with regard to	about, in

Similarly, avoid the use of adjectives and nouns that are formed from verbs. If the root verbs are used instead, the writing will be more concise and more vivid. For example,

Noun form	Verb form
We made an adjustment on our books	We adjusted our books
We are sorry we cannot make a replacement of	We are sorry we cannot replace
Please make a correction in our order	Please correct our order

Be on the lookout for unnecessary adjectives and adverbs which tend to clutter letters without adding information or improving style. Such unnecessary words tend to distract the reader and make it more difficult for him to grasp the main points. Observe how the superfluous words, italicized in the following example, obscure the meaning: "You may be *very much* disappointed to learn that the *excessively large* demand for our *highly popular recent* publication, 'Your Income Taxes,' has led to an *unexpected* shortage of this *attractive* publication and we *sadly* expect they will not be replenished until *quite* late this year."

Summarizing, then, *a good letter is simple and clear, with short, simple words, sentences, and paragraphs. Related parts of sentences and paragraphs are kept together and placed in an order which makes it easy for the reader to follow the main thoughts.*

Be Natural. Whenever possible, use a human touch. Use names and personal pronouns to let the reader know the letter was written by a person, not an institution. Instead of saying, "It is the policy of this agency to contact its clients once each year to confirm their status," try this: "Our policy, Mr. Jones, is to confirm your status once each year."

Use Concrete Nouns. Avoid using abstract words and generalizations. Use names of objects, places, and persons rather than abstractions.

Use Active Verbs. The passive voice gives a motionless, weak tone to most writing. Instead of "The minutes were taken by Mrs. Smith," say, "Mrs. Smith took the minutes." Instead of "The plans were prepared by the banquet committee," say, "The banquet committee prepared the plans."

Use a Natural Tone. Many people tend to become hard, cold, and unnatural the moment they write a letter. *Communicating by letter should have the same natural tone of conversation used in everyday speech.* One way to achieve a natural and personal tone in the majority of letters is through the use of personal pronouns. Instead of saying, "Referring to your letter of March 5, reporting the non-receipt of goods ordered last February 15, please be advised that the goods were shipped as requested," say, "I am sorry to hear that you failed to receive the items you ordered last February 15. We shipped them the same day we received your letter."

Forms

In most businesses and public service agencies, repetitive work is simplified by the use of *forms*. Forms exist for nearly every purpose imaginable: for ordering supplies, preparing invoices, applying for jobs, applying for insurance, paying taxes, recording inventories, and so on.

While the forms encountered in different agencies may differ widely, several principles should be applied in completing any form:

- *Legibility*. Entries on forms should be clear and legible. Print or type wherever possible. When space provided is insufficient, attach a supplementary sheet to the form.
- *Completeness*. Make an entry in every space provided on the form. If a particular space does not apply to the applicant, enter there the term "N/A" (for "not applicable"). The reader of the completed form will then know that the applicant did not simply overlook that space.
- *Conciseness*. Forms are intended to elicit a maximum amount of information in the least possible space. When completing a form, it is usually not necessary to write complete sentences. Provide the necessary information in the least possible words.
- *Accuracy*. Be sure the information provided on the form is accurate. If the entry is a number, such as a social security number or an address, double-check the correctness of the number. Be sure of the spelling of names, No one appreciates receiving a communication in which his name is misspelled.

Memoranda

The written communications passing between offices or departments are usually transmitted in a form known as *"interoffice memorandum."* The headings most often used on such "memos" are:

- TO: identifying the addressee,
- FROM: identifying the sender or the originating office,
- SUBJECT: identifying briefly the subject of the memo,
- DATE: identifying the date the memo was prepared.

Larger agencies may also use headings such as FILE or REFERENCE NO. to aid in filing and retrieving memoranda.

In writing a memo, many of the same rules for letter-writing may be applied. Both the appearance and tone of the memo should create a pleasing impression. The format should be neat and follow the standards set by the originating office. The tone should be friendly, courteous, and considerate. The language should be clear, concise, and complete.

Memos usually dispense with salutations, complimentary closings, and signatures of the writers. In most other respects, however, the memorandum will follow the rules of good letter-writing.

Minutes of Meetings

Most formal public-service organization conduct meetings from time to time at which group decisions are made about agency policies, procedures, and work assignments. The records of such meetings are called *minutes*.

Minutes should be written as clearly and simply as possible, summarizing only the essential facts and decisions made at the meeting. While some issue may have been discussed at great length, only the final decision or resolution made of it should be recorded in the minutes. Information of this sort is usually included:

- Time and place of the call to order,
- Presiding officer and secretary,
- Voting members present (with names, if a small organization),
- Approval and corrections of previous minutes,
- Urgent business,
- Old business,
- New business,
- Time of adjournment,
- Signature of recorder.

Minutes should be written in a factual and objective style. The opinions of the recorder should not be in evidence. Every item of business coming up before a meeting should be included in the minutes, together with its disposition. For example:

- "M/S/P (Moved, seconded, passed) that Mr. Thomas Jones take responsibility for rewriting the personnel procedures manual."
- "Discussion of the summer vacation schedule was tabled until the next meeting."
- "M/S/P, a resolution that no client of the agency should be kept waiting more than 20 minutes for an interview."

Note that considerable discussion may have surrounded each of the above items in the minutes, but that only the topic and its resolution are recorded.

Short Reports

The public-service employee often is called upon to prepare a short report gathering and interpreting information on a single topic. Reports of this kind are sometimes prepared so that all the relevant information may be assembled in one place to aid the organization in making certain decisions. Such reports may be read primarily by the staff of the organization or by others closely related to the decision-making process.

Reports may be prepared at other times for distribution to the public or to other agencies and institutions. These reports may serve the purpose of informing public opinion or persuading others on matters of public policy.

Whatever the purpose of the short report, its physical appearance and style of presentation should be designed to create a favorable impression on the reader. Even if the report is distributed only within the writer's own unit, an attractive, clear, thorough report will reflect the writer's dedication to his assignment and the pride he takes in his work.

Some guidelines which will assist the trainee in preparation of effective short reports include use of the following:

- A good quality paper;
- Wide and even margins, allowing binding room;
- An accepted standard style of typing;
- A title page;
- A table of contents (for more lengthy reports only);
- A graphic numbering or outlining system, if needed for clarity;
- Graphics and photos to clarify meaning when useful;
- Footnotes, used sparingly, and only when they contribute to the report;
- A bibliography of sources, using a standard citation style.

A discussion of the organization of content for informational reports follows later in this document.

Telegrams and Cables

From time to time messages of special urgency must be sent by public telegraph wires, cables, and radio. With this service, written communications may be exchanged worldwide within minutes at a cost not greatly more than for a letter.

The public-service employee should be familiar with the telegraph service and able to prepare written messages for this medium. The student should be familiar with the classes of service available: "full-rate service," "day letter," and "night letter," since the class of service will affect the style of writing.

Skill in preparing telegraph messages rests largely on the writer's ability to summarize. The essential information must be presented in the fewest possible words. Good messages would follow these guidelines:

- Omission of articles and prepositions unless essential to meaning,
- Use of verb forms of the fewest words,
- Use of single words rather than phrases,
- Omission of unnecessary information and words.

For example:

> "I am taking American Airlines Flight 222 from Chicago at 8:15 Wednesday evening and will arrive at Los Angeles International Airport at 10:15. I would appreciate it if you would meet me."

Can be compressed to:

> "Arriving American 222 Los Angeles International Wednesday 10:15 PM. Please meet."

The minimum charge on a full-rate telegram is based upon 15 words. The student should develop skill in writing 15-word summaries in telegraph style.

News Releases

From time to time, the public-service employees may be called upon to prepare a news release for his agency. Whenever the activities of the agency are newsworthy or of interest to the public, the agency has an obligation to report such activities to the press. The most common means for such reporting is by using the press release. Most newspapers and broadcasting stations are initially informed of agencies' activities by news releases distributed by the agencies themselves. Thus, the news release is a basic tool for communicating with the public served by the agency.

The news release is written in news style, with these basic characteristics:

- Sentences are short and simple.
- Paragraphs are short (one or two sentences) and relate to a single item of information.
- Paragraphs are arranged in *inverted order* - the most important in information appears first.
- The first or *lead* paragraph summarizes the entire story. If the reader went no further, he would have the essential information.
- Subsequent paragraphs provide further details, the most important occurring first.
- Reported information is attributed to sources; that is, the source of the news is reported in the story.
- The expression of the writer's opinions is scrupulously avoided.
- The 5 W's (who, what, why, where, when) are included.

News releases should be typed double spaced on standard 8½ x 11 paper, with generous margins and at least 2" of open space above the lead paragraph. Do not write headlines - that is the editor's job. At the top of the first page of the release include the name of the agency releasing the story and the name and phone number of the person to contact if more information is needed. If the release runs more than one page, end each page with the word "-more-" to indicate that more copy follows. End the release with the symbols "###" to indicate that the copy ends at that point.

Accuracy and physical appearance are essential characteristics of the news release. Typographical errors, or errors of fact, such as misspelled names, lead editors to doubt the reliability of the story. Great care should be taken to assure the accuracy and reliability of a news release.

2. REPORTING ON A TOPIC

At one time or another, most public-service employees will be asked to prepare a report on some topic. Usually the need for the report grows out of some policy decision contemplated by the agency for which full information must be considered. For example:

- Should the agency undertake some new project or service?
- Should working conditions be changed?
- Are new specialists needed on the staff?
- Or should a branch office be opened up?

Or any of a hundred other such decisions which the agency must make from time to time.

When called upon to prepare such a report, the employee should have a model to follow which will guide his collection of information and will help him to prepare an effective and useful report.

As with other forms of written communication, both the physical appearance and content of the report are important to create a favorable impression and to engender confidence. The physical appearance of such reports has been discussed earlier; additional suggestions for reports are given in Unit 3. Basic guidelines follow below for organizing and preparing the content.

Preparation for the Report

What is the Purpose of the Report? The preparer of the report should have clearly in mind why the report is needed:

- What is the decision being contemplated by the agency?
- To what use will the report be put?

Before beginning to prepare the report, the writer should discuss its purpose fully with the decision-making staff to articulate the purpose the report is intended to serve. If the employee is himself initiating the report, it would be well to discuss its purpose with colleagues to assure that its purpose is clear in his own mind.

What Questions Should the Report Answer? Once the purpose of the report is clear, the questions the report must answer may begin to become clear. For example, if the decision faced by the agency is whether or not to offer a new service, questions may be asked such as these:

- What persons would be served by the new service?
- What would the new service cost?
- What new staff would be needed?
- What new equipment and facilities would be needed?
- What alternative ways exist for offering the service?
- How might the new service be administered?

And so on. Unless the purpose of the report is clear, it is difficult to decide what specific questions need to be answered. Once the purpose is clear, these questions can be specified.

Where Can the Relevant Information be Obtained? Once the questions are clear in the writer's mind, he can identify the information he will need to answer them. Information may usually be obtained from two general sources:

- *Relevant documents.* Records, publications, and other reports are often useful in locating the information needed to answer particular questions. These may be in the files of the writer's own agency, in other agencies, or in libraries.

- *Personal contacts.* Persons in a position to know the needed information may be contacted in person, by phone, or by letter. Such contacts are especially important in obtaining firsthand accounts of previous experience.

The Text of the Report

What are the Answers to the Questions? Once the relevant information is in hand, the answers to the questions may be assembled.

- What does the information reveal? This activity amounts to summarizing the information obtained. It often helps to organize this summary around the specific questions asked by the report. For example, if the report asks in one part, "What are the costs of the new service likely to be?" one section of the report should summarize the information gathered to answer this question.

Organizing the Report. The organization of a report into main and sub-sections depends upon the nature of the report. Reports will differ widely in their organization and treatment. In general, however, the report should generally follow the pattern previously discussed. That is, reports which generally include the following subjects in order will be found to be clear in their intent and to communicate effectively:

- *Description of problem or purpose.* Example: "One problem facing our agency is whether or not we should extend our hours of operation to better serve the public. This report is intended to examine the problem and make recommendations."

- *Questions to be answered.* Example: "In examining this problem, answers were sought to the following questions: What persons would be served? What would it cost? What staff would be needed?"

- *Information sources.* Example: "To answer these questions, letters of complaint for the past three years were examined. Interviews with clients were conducted by phone and in person, phone interviews were conducted with the agency directors in Memphis, Philadelphia, and Chicago."

- *Summary of findings.* Example: "At least 25 percent of the agency's clients would be served better by evening or Saturday service. The costs of operating eight hours of extended service would be negligible, since the service could be provided by rescheduling work assignments. The present staff report they would be inconvenienced by evening and Saturday work assignments."

The Writer's Responsibilities. It is the writer's responsibility to address finally the original purpose of the report. Once the questions have been answered, an informed judgment can be made as to the decision facing the agency. It is at this stage that the writer attempts to draw conclusions from the information he has gathered and summarized. For example, if the original purpose of the report was to help make a decision about whether or not the agency should offer a new service, the writer should draw conclusions from the information and recommend either for or against the new service.

Conclusions and Recommendations. Example: "It appears that operating during extended hours would better serve a significant number of clients. The writer recommends that the agency offer this new service. The present staff should be given temporary assignments to cover the extended hours. As new staff are hired to replace separating persons, they should be hired specifically to cover the extended hours."

3. PERSUASIVE WRITING

Often in life, people are called upon to persuade individuals and groups to adopt ideas believed to be good, or attitudes favorable to ideas thought to be worthwhile or behavior believed to be beneficial. The public service employee may find he must persuade the staff of his own agency, his superiors, the clients of the agency, or the general public in his community.

Persuading others by means of written and other forms of communication is a difficult task and requires much practice. Some principles have emerged from the study of persuasion which may provide some guidelines for developing a model for persuasive writing.

General Guidelines for Writing Persuasively

Know the Credibility of the Source. People are more likely to be persuaded by a message they perceive originates from a trustworthy source. Their trust is enhanced if the source is seen as authoritative, or knowledgeable on the issue discussed in the message. Their trust is increased also if the source appears to have nothing to gain either way, has no vested interest in the final decision. Then, the assertions made in persuasive writing should be backed up by referencing trustworthy and disinterested information sources.

Avoid Overemotional Appeals. Appealing to the common emotions of man - love, hate, fear, sex, etc. - can have a favorable effect on the outcome of a persuasive message. But care should be taken because, if the appeal is too strong, it can lead to a reverse effect. For example, if an agency wanted to persuade the public to get chest X-rays, it would have much greater chance of success if it adopted a positive and helpful attitude rather than trying to frighten them into this action. For instance, appealing mildly to the sense of well-being which accompanies knowledge of one's own good health, instead of shocking the public by showing horror pictures of patients who died from lack of timely X-rays.

Consider the Other Man's Point of View. To persuade another to one's own point of view, should the writer include information and arguments contrary to his own position? Or should he argue only for his own side?

Generally, it depends on where most of the audience stand in the first place. If most of the audience already favor the position being advocated, then the writer will probably do better including only information favorable to his position. However, if the greater part of the audience are likely to oppose this position, then the writer would probably be better off including their arguments also. In this case, he may be helping his cause by rebutting the opposing arguments as he introduces them into the writing.

An example of this technique might occur in arguing for such an idea as a four-day, forty-hour workweek. Thus: "Many people feel that the ten-hour day is too long and that they would arrive home too late for their regular dinner hour. But think! If you have dinner a littler later each night, you'll have a three-day weekend every week. More days free to go fishing, or camping. More days with your wife and children." That is good persuasive writing!

Interpersonal Communications

The important role of interpersonal communication in persuading others - face-to-face and person-to-person communications - has been well documented. Mass mailings or printed messages will likely have less effect than personal letters and conversations between persons already known to each other. In any persuasion campaign the personal touch is very important.

An individual in persuading a large number of persons will likely be more effective if he can organize a letter-writing campaign of persuasive messages written by persons favorable to his position to their friends and acquaintances, than if his campaign is based upon sending out a mass mailing of a printed message.

Conditions for Persuading. In order for an audience of one or many to be persuaded in the manner desired, these conditions must be met:

° the audience must be *exposed* to the message,

° members of the audience must *perceive* the intent of the message,

° they must *remember* the message afterwards,

° each member must *decide* whether or not to adopt the ideas.

Each member of the audience will respond to a message differently. While every person may receive the message, not everyone will read it. Even among those who read it, not everyone will perceive it in the same way. Some will remember it longer than others. Not everyone will decide to adopt the ideas. These effects are called *selective exposure, selective perception, selective retention,* and *selective decision.*

The Persuasion Campaign. How can one counteract these selective effects in persuading others? One thing that is known is that *people tend to be influenced by persuasive messages which they are already predisposed to accept.* This means a person is more likely to persuade people a little than to persuade them a lot.

In planning a persuasion campaign, therefore, the messages should be tailored to the audiences. Success will be more likely if one starts with people who believe *almost* as the writer wants to persuade them to believe - people who are most likely to agree with the position advocated.

The writer also wants to use arguments based on values the particular audience already accepts. For example, in advocating a new teen-age job program, he might argue with business men that the program will help business; with parents, that it will build character; with teachers, that it is educational; with taxpayers, that it will reduce future taxes; and so on.

The idea is to find some way to make sure that each member of the particular audiences reached can see an advantage for himself, and for the writer to then tailor the messages for those audiences.

4. INSTRUCTIONAL WRITING

Another task that the public-service employee may expect to face from time to time is the instruction of some other person in the performance of a task. This may sometimes involve preparing written instructions to other employees in the unit, or preparing a training manual for new employees.

It may sometimes involve preparing instructional manuals for clients of the unit, such as "How to Apply for a Real Estate License," "How to Bathe your Baby," or "How to Recognize the Symptoms of Heart Disease."

Whatever the purpose or the audience, certain principles of instruction may be applied which will help make more effective these instructional or training communications. These are: *advance organizers, practice, errorless learning,* and *feedback.*

Advance Organizers

At or near the beginning of an instructional communication, it helps the learner if he is provided with what can be called an "advance organizer." This element of the communication performs two functions:

- it provides a framework or "map" for the leader to organize the information he will encounter,
- it helps the learner perceive his purpose in learning the tasks which will follow.

The first paragraphs in this section, for example, serve together as an advance organizer. The trainee is informed that he may be called upon to perform these tasks in his job *(perceived purpose),* and that he will be instructed in advance organizers, practice, errorless learning, and feedback *(framework, or "map").*

Practice

The notion of *practice makes perfect* is a sound instructional principle. When trying to teach someone to perform a task by means of written communication, the writer should build in many opportunities for practicing the task, or parts of it. This built-in practice should be both appropriate and active:

- *Appropriate practice* is practice which is directly related to learning the tasks at hand.
- *Active practice* is practice in actually performing the task at hand or parts of it, rather than simply reading about the task, or thinking about it.

By inserting questions into the text of the communication, by giving practice quizzes, exercises, or field work, one can build into his instructional communication the kind of practice necessary for the reader to readily learn the task.

Errorless Learning

The practice given learners should be easy to do. That is, they should not be asked to practice a task if they are likely to make a lot of mistakes. When a mistake is practiced it is likely to recur again and again, like spelling "demons," which have been spelled wrong so often it's difficult to recall the way they should be spelled. Because it is better to practice a task right from the first, it is important that learners do not make errors in practice.

- One method for encouraging correct practice is to give the reader hints, or *prompts*, to help him practice correctly.
- Another method is to instruct him in a logical sequence a little bit at a time. Don't try to teach everything at once. Break the task down into small parts and teach each part of the task in order. Then give the learner practice in each part of the task before giving him practice in the whole thing.
- A third way of encouraging errorless learning is to build in practice and review throughout the communication. The learner may forget part of the task if the teacher doesn't review it with him from time to time.

Remember, people primarily learn from what they do, so build in to the instructional communication many opportunities for the learner to practice correctly all of the parts of the task required for learning, first separately and then all together.

Feedback

The reader, or learner, can't judge how well he is learning the task unless he is informed of it. In a classroom situation, the teacher usually confirms that the learner has been successful, or points out the errors he made, and provides additional instruction. An instructional communication can also help learners in the same way, by providing *feedback* to the learner.

Following practice, the writer should include in his instructional communication information which will let the reader know whether he performed the task correctly. In case he didn't, the writer should also include some further information which will help the reader perform it correctly next time. This feedback, then, performs two functions:

- it helps the learner confirm that his practice was done correctly, and
- it helps him correct his performance of the task in case he made any errors.

Feedback will be most helpful to the learner if it occurs immediately following practice. The learner should be brought to know of his success or his errors just as soon as possible after practice.

STUDENT LEARNING ACTIVITIES

- Write "asking" and "answering" letters, and answer a letter of complaint, using the format assigned by the teacher.
- Write memoranda to other "offices" in a fictitious organization. Plan a field trip using only memos to communicate with other students in the class.
- Take minutes of a small group meeting. Or attend a meeting of the school board and take minutes.
- Write a short report on a public service occupation of special interest to you.
- Write a 15-word telegram reserving a single room at a hotel and asking to be picked up at the airport.
- Write a news release announcing a new service offered to the public by your agency.
- Based upon hearing a reading or pretaping of a report, summarize the report in news style.
- View films on effective communication, for example, *Getting the Facts*, *Words that Don't Inform*, and *A Message to No One*.
- For a given problem or purpose, compile a list of specific questions you would need to answer to write a report on the topic.

- For a given list of questions, discuss and compile a list of information sources relevant to the questions.
- As a member of a group, consider the problem of "What field trip should the class take to help students learn how to write an effective news release?" What questions will you need to answer? Where will you obtain your information?
- As a member of a group, gather the information and prepare a short report based on it for presentation to the class.
- Write a report on a problem assigned by your teacher.
- Write a brief persuasive letter to a friend on a given topic. Assume he does not already agree with you. Apply principles of source credibility, emotional appeals, and one or both sides of the issue to persuade him.
- Plan a persuasive campaign to persuade a given segment of your community to take some given action.
- Write a short instructional communication on a verbal learning task assigned by your teacher.
- Write a short instructional communication on a learning task which involves the operation of equipment.
- Try your instructional communications with a fellow student to check for errors during practice.

TEACHER MANAGEMENT ACTIVITIES

- Have students practice letter writing. Assign letters of "asking" and "answering." Read them a letter of complaint and ask them to write an answering letter. Establish common rules of format and style for each assignment. Change the rules from time to time to give practice in several styles.
- Have small groups plan an event, such as a field trip, assigning the various tasks to one another using <u>only</u> memoranda. Evaluate the effectiveness of each group's memo writing by the speed and completeness of their planning.
- Have the class attend a public meeting. Assign each the task of taking the minutes. Evaluate the minutes for brevity and completeness.
- Encourage each student to prepare a short report on a public service occupation of special interest to himself.
- Give the students practice in writing 15-word telegrams.
- Have the students prepare a news release announcing some new service offered to the public, such as "Taxpayers can now obtain help from the Internal Revenue Service in completing their income tax forms as a result of a new service now being offered by the agency."
- Give the students practice in summarizing and writing leads

by giving them the facts of a news event and asking them to write a one or two-sentence lead summarizing the significant facts of the event.

- Read a speech or a story. Have students write a summary and a report of the speech or story in news style.
- Show films on effective communication, for example, *Getting the Facts, Words that Don't Inform,* and *A Message to No One*.
- State a general problem and have each student prepare a list of the specific questions implied by the problem.
- State a list of specific questions and discuss with the class the sources of information which might bear upon each of the questions.
- Have small groups consider and write short reports jointly on the general problem, "What field trip should the class take to help students learn how to write an effective news release?" Have each group identify the specific questions to be answered, with sources for needed information.
- Have each student identify and prepare a short report on a general problem of interest.
- Assign students to work in groups of three or four to draft a letter to a friend to persuade him to make a contribution to establish a new city art museum.
- Assign the students to groups of five or six, each group to map out a persuasive campaign on a given topic. Some topics are "Give Blood," "Get Chest X-Ray," "Quit Smoking," "Don't Litter," "Inspect Your House Wiring," etc.
- Have each student identify a simple verbal learning task and prepare an instructional communication to teach that task to another student not familiar with the task.
- Have each student prepare an instructional manual designed to train someone to operate some simple piece of equipment, such as an adding machine, a slide projector, a tape recorder, or something of similar complexity.
- Have each student try his instructional communication out on another student, unfamiliar with the task. He should observe the activities and responses of the trial student to identify errors made in practice. He should revise the communication, adding practice, review, and prompts wherever needed to reduce errors in practice.

EVALUATION QUESTIONS

Written Communications

1. **Which type of letter would be correct for a public service worker to send:**
   A. A letter containing erasures ______
   B. A letter reflecting goodwill ______
   C. A rude letter ______
   D. An impersonal letter ______

2. **Memos usually leave out:**
   A. Complimentary closings ______
   B. The name of the sender ______
   C. The name of the addressee ______
   D. The date the memo was sent ______

3. **A good business letter would not contain:**
   A. Short, simple words, sentences, and paragraphs ______
   B. Information contained in the letter being answered ______
   C. Concrete nouns and active verbs ______
   D. Orderly placed paragraphs ______

4. **In writing business letters it is important to:**
   A. Use a conversational tone ______
   B. Use a hard, cold tone ______
   C. Use abstract words ______
   D. Use a passive tone ______

5. **Messages between departments in an agency are usually sent by:**
   A. Letter ______
   B. Memo ______
   C. Telegram ______
   D. Long reports ______

6. **Repetitive work can be simplified by the use of:**
   A. Memos ______
   B. Telegrams ______
   C. Forms ______
   D. Reports ______

7. **In filling out forms and applications, it is important to be:**
   A. Legible ______
   B. Complete ______
   C. Accurate ______
   D. All of the above ______

8. **Memos should be:**
   A. Clear
   B. Brief
   C. Complete
   D. All of the above

9. **Minutes of meetings should not include:**
   A. The opinions of the recorder
   B. The approval of previous minutes
   C. The corrections of previous minutes
   D. The voting members present

10. **Reports are written by public service workers to:**
   A. Assemble information in one place
   B. Aid the organization in making decisions
   C. Inform the public and other agencies
   D. All of the above

11. **News releases should include:**
   A. A lead paragraph summarizing the story
   B. Long paragraphs about many topics
   C. The writer's opinion
   D. All of the above

12. **Readers of news releases and reports are influenced by the:**
   A. Content of the material
   B. Accuracy of the material
   C. Physical appearance of the material
   D. All of the above

13. **The contents of a report should not include:**
   A. A description of the problem
   B. The questions to be answered
   C. Unimportant information
   D. A summary of findings

14. **People tend to be influenced easier if:**
   A. They can see something in the position that would be advantageous to them
   C. They are almost ready to agree anyhow
   C. The appeal to the emotions is not overly strong
   D. All of the above

Answer Key

1. B	4. A	7. D	10. D	13. C
2. A	5. B	8. D	11. A	14. D
3. B	6. C	9. A	12. D	

BASIC FUNDAMENTALS OF REPORT WRITING

CONTENTS

BASIC FUNDAMENTALS OF REPORT WRITING

INSTRUCTIONAL OBJECTIVES

1. Ability to understand how reports are used.
2. Ability to form an appreciation for the importance of effective report writing in public service occupations.
3. Ability to define and differentiate between the types of reports.
4. Ability to select the proper form for a report.
5. Ability to recognize the steps in preparing a good report.
6. Ability to demonstrate a knowledge of correct word usage and its importance in a report.
7. Ability to demonstrate the proper grammatical structure in report writing.
8. Ability to explain the importance of using a dictionary in report writing.

CONTENT

INTRODUCTION

A report is the communication of information in the most convenient form to someone who wants or needs it. It is difficult to overstate the importance of reports in our complex society today. They provide the data on which far-reaching decisions and policy are determined. A report that is not accurate or that is incomplete may cause the writer and the receiver misunderstanding, antagonism, and costly errors. Repeated errors in reports may well cost the writer his position.

Many of the reports that public service workers make are oral ones. However, the more important reports are always required to be in writing.

A report is always addressed to someone - the audience. The audience is usually a superior who needs the information to determine what action to take or what recommendations to pass on to his superior. The good report writer analyzes the audience to which the report will be directed.

° Should the report be a formal document or should the report be an informal memo?

° How will the audience use the report?

° Will it be passed along to others in the organization, or people in other areas?

The report writer's answers to questions such as these may well determine the format that is selected for the report.

1. TYPES OF REPORTS

The Formal Report. Formal reports usually follow a format prescribed by the agency or department. A central office typically supplies the form on which the report is to be prepared and submitted. A copy of a previous report is frequently used as a guideline in preparing the new one. This procedure saves considerable time and makes it easier for the central office to analyze and interpret the information contained in the report.

The Informal Report. Another type of report, the informal one, frequently takes the form of a note or memo requesting additional supplies, services, or providing suggestions for the supervisor or a fellow worker. The informal report could possibly contain abbreviations, first names, and technical jargon that would be meaningless to anyone not acquainted with the situation in which it was written.

The Semi-Formal Report. A semi-formal report fits between the casual or informal memo and the highly structured and authoritative formal report. The semi-formal report is to be considered more important than the informal one and consequently is written with more thought and care than the informal report.

2. EXAMPLES OF REPORTS

School Setting. The various types of reports may be illustrated in school settings:

- *The Formal Report.* Progress reports, attendance reports, accident reports, and classroom reports prepared by students; these are all examples of well-structured formal reports. The importance of such reports is great. For example, the attendance reports determine how many teachers will be added or deleted from the faculty, in addition to being the basis for computing the state's financial contribution to the local school district.

- *The Semi-Formal Report* is illustrated by the office referral requesting action because of a student's behavior. It may result in a student-administrator conference, a parent conference, or even student suspension from school.

- *The Informal Report.* The note that a teacher leaves in the school mail for the staff assistant in charge of audiovisual equipment requesting service on the malfunctioning movie projector is an example of an informal report.

Public Service Agencies. The public-service report writer will find that he is writing:

- About specific subjects, and about various technicalities or events associated with the governmental agency or a responsibility assigned to it.
- With certain formal elements, such as the technical vocabulary of the department and the use of forms prepared by the department or agency.
- With an attitude of impartiality and objectivity, taking extreme care to convey information accurately and concisely, and with no attempt to arouse emotions.

Police Reporting - An illustration which would be meaningful to students may be made by examining what a policeman would include in a report when investigating an automobile accident. The policeman's report must contain such specific facts as:

- The names, addresses, phone numbers, ages, drivers' license numbers, and the identification of witnesses to the accident.
- Where the accident occurred - address.
- Description of the accident scene.
- The witnesses' observations - obtaining not only their account of events, but from what point they observed the event, and what they were doing at the time.
- Time and day of the accident.
- Weather conditions at the time of accident.

- The location of traffic signals, stop signs, skid marks, traffic lanes, etc.
- Description of the injuries, if any, or lack of apparent injuries.
- Description of the automobiles involved and apparent damages.

The investigating officer will have a pad of accident report forms which will outline the information needed. This insures that in the emotion and tension of the accident scene the investigating officer will be reminded of the information that he needs to include in the report. Numbers will be used to code certain information. This speeds up the writing, simplifies the recording and filing of the information, and makes the report less emotional.

The accident report will be of major importance when the insurance companies settle the claim. If there is a dispute over the settlement, the police report will be entered into the court record. The officer who wrote the report will also have to testify. The testimony may not be called for years after the accident and the officer will have to use the report to refresh his memory. It would probably be difficult for the officer otherwise to recall all of the events or even to notice them all at the time, as he may have had to direct traffic, administer first aid, call an ambulance, and keep curious bystanders out of the way. Even though human life may have been lost or saved by the policeman's actions on the scene, the primary issues in which the court or insurance adjusters will be interested are the facts in his report.

Other Reporting Areas - There are other workers in the public service area, such as teachers' aides, recreation aides, and forestry aides, who are likely to be required to complete accident reports while on the job. While their reports may not be as sensational as the police report, they are important and may be used in subsequent litigation.

It should also be pointed out that any report may have far-reaching ramifications. However, it is wise to be especially careful if a financial settlement between two adversaries will be affected by the report.

3. THE REPORT WRITER

Regardless of the type of report that is being prepared, the key to a good report is a good writer. The writer is the prime mover of the entire process. He gathers the material, interprets it, organizes it, and chooses the vocabulary. The writer imposes meaning merely by determining whether to make the report a formal, a semi-formal, or an informal one.

The reader is led to an understanding by an orderly presentation of the facts. The facts are determined by the writer's examination and observations of the situation or activity. The report's validity hinges upon the competence and objectivity of the writer and the accuracy of his observations.

Competence in report writing means that an individual has acquired the technique of careful inquiry. This may be illustrated by the police officer at the scene of the accident. He will take statements from witnesses, take measurements of skid marks, inspect and record the automobiles' registrations, and inspect the scene of the accident for real evidence that may have been instrumental in the cause of the accident, such as flat tires, conditions of the tires on the automobiles, mechanical defects, health of the drivers, etc. The report writer must acquire the habit of orderly thinking from problems to solutions, and learn to write an exact statement.

A good report communicates clear ideas in simple language. The organization is orderly and fast moving. It is easy to read and uses visual aids where they will best carry forward the message. These skills are required of almost every member of governmental agencies. Management quickly recognizes competence in the preparation of reports. The individual who writes them well will assume a position of leadership at every level where accurate data supplies the basis for a difficult decision.

4. CHOOSING BEST WORDS

A major aspect of clear and concise report writing is being careful to select the proper word to express the thought exactly.

Use Proper Word for Intended Meaning. Since any one word may have a number of different meanings in different contexts, the writer must be certain that the word chosen is appropriate to convey the intended meaning within the human situation in which it is used. Paul Douglas, in his book *Communicating Through Reports*, illustrates this point with "word." He lists twelve different meanings, ranging from the preacher expounding the *Word* (scripture) to the wife's having the *last word* (decision). This relationship of meaning to the environment and circumstances of its use is known as *context*.

Nearly all words mean more than they seem to mean; they possess associative meanings, almost outlying areas of suggestive values. The bare dictionary meaning of a word is its *denotation*. The *connotation* of a word relates to the suggestions and associations which have surrounded it.

For example, a dictionary definition of the word "gold" is "a precious yellow metal which is highly malleable and ductible. This is its denotation. But along with its denotative meaning, gold has also been associated with wealth, value, color, and power. These characteristics are the connotations of gold.

Beyond the core of the dictionary definition are suggestions, associations, and implications. Connotations must be watched as they have practical consequences for the report writer. Words and phrases with unpleasant connotations may become fighting words which blind the reader to anything else in the report. The following words are a few that have negative suggestions: mistakes, inefficient, death, refuse, error, rejecting, prohibited, unfair, poor judgment, and failed.

The writer may avoid irritation by choosing words with positive or neutral connotations. Note the difference in meaning of the following words:

- car, automobile, limousine
- inebriated, intoxicated, drunk
- portly, stout, obese
- smell, smell bad, stink
- slow, lazy, sluggish
- firm, obstinate, pigheaded
- Negro, colored man, black man
- dog, hound, mongrel

To use words both accurately and effectively, one must understand their connotations as well as their denotations. Social welfare aides, employment service workers, and probation services aides need to be especially alert to those words which are unique to the groups of people they serve. When wishing to express something with literal accuracy, the report writer will rely wholly upon the standards set by the department or dictionary definitions.

Eliminate Slang in Reports. *Slang should not be used in reports except when reporting dialogue.* There are several reasons for not using slang:

- First, many slang words and expressions are so short-lived that they will be outmoded before the report will have lived its useful life.

- Second, the use of slang expressions may be an excuse for not searching for the exact words to express the meaning.

Many slang expressions are only rubber stamps; to refer to a person as a "dude" hardly expresses exactly any critical judgment or intelligent description. To argue that such a word conveys precisely the intended meaning is to reveal a poverty of vocabulary, or careless thinking and laxness. The most serious charge against slang is that it becomes a substitute for thinking. Public service workers are likely to encounter the following slang:

- grubby
- get burned
- the boss
- real cool dude
- far out
- right on
- groovy
- bummer
- a drag
- rip off
- split
- low rider
- doing your own thing
- busted

The list could go on indefinitely. It would be very difficult to get a widely acceptable definition for any one of the above slang words. *Slang does express feeling, but when writing reports, the main concern is facts, not feelings.*

5. CHOOSING BEST PHRASES

Don't Use Weasel Words. Other words to avoid in writing reports are what some authors call "weasel words." *A weasel word is a phrase that rids the writer of any responsibility for a statement.* By the use of the weasel word, the writer attempts to wriggle out of a position of accountability for an observation, inference, or statement, as may be seen by these examples:

- It would be difficult to estimate . . .
- It is too early to say whether . . .
- It is generally believed that . . .
- It would appear that . . .
- There may be a tendency toward . . .

Weasel words are usually used with the passive voice, which is discussed later in this unit under "Writing Sentences."

Clichés to Avoid. Public-service workers, like many other report writers, too often use words and phrases which are trite, outworn, commonplace, and flat stereotypes. Here are some examples of cliches which should be avoided:

- It certainly merits study . . .
- We will see what may be done . . .

- This suggestion certainly has merit . . .
- The matter is receiving our closest attention . . .
- We will explore every avenue . . .
- The handwriting on the wall . . .
- Let's get down to brass tacks . . .
- Naturally, the child's interest is our concern . . .
- Fools rush in . . .
- In the last analysis . . .
- No thinking man . . .
- The skeleton in the closet . . .
- Let sleeping dogs . . .

Cliches are similar to slang in that they are but rubber stamps, "stereotyped plates" of thought and expression. They save the writer the trouble of thinking exactly what he means. Consequently, they get in the way of clear and concise report writing.

Don't Use Excess Wordage. Diction, to be effective, must be as economical as possible. Necessary words should not be omitted nor should the report be sketchy. However, wordiness should be avoided as it lessens the force of expression. *In forceful writing, the ratio of ideas to words is high.* Conciseness alone will not achieve effective writing, but it is extremely difficult to write forcefully if you use two or three words to convey the idea which one word would express. Note these examples of excess or "superfluous" words:

- absolutely parallel
- first beginnings
- free gratis
- someone I met up with
- perfect circle
- the sunset in the west
- resume again
- join together

Don't be Pompous. When it is possible to reduce a group of words to a single word, it should be done. Here are some examples of economy in diction that are achieved by reducing pompous phrases to more natural words:

Pompous	Natural
in the nature of	like
for the purpose of	for
prior to	before
subsequent to	after
despite the fact that	though
give consideration to	consider
give instruction to	instruct

- is due in large measure to
- confidential nature

- is due largely to
- confidential information

6. WRITING SENTENCES

Be Simple. *One of the most important rules to follow in writing a report is to keep your sentences short, usually twenty words or less.* Language, like a machine, will be the most efficient when it operates on the principle that the more simple and better arranged the parts, the greater the effect which is produced. The subject-verb-object sentence is the best arrangement of parts in the simple report.

Keep Active. Verbs should be kept active. Avoid the passive voice; this puts excess words in a sentence, and its dullness derives as much from its extra wordage as from its impersonality. John was hit by Don says no more than Don hit John but takes 66 percent more words.

The passive voice's inevitable "was" and "by" do nothing but connect; worse, all the "was's" and "by's" and "has been's" actually get in the way of words carrying the meaning. It's like underbrush, it slows you down and hides what the report reader should see.

The passive voice, in its wordiness, is likely to be unclear even on the surface. When it eliminates the subject of the verb, as it usually does, it is intrinsically unclear. For example, a sentence written in the passive voice will often begin: "This evidence has been selected because...." The reader cannot tell who did the selecting. Does the writer mean that he picked it or does he describe some process of popular selection? It is usually surmised that the writer did the selecting, but why doesn't he say so and save a word and avoid confusion? "I selected this evidence because...." *The report writer should be careful not to leave the reader with any assumptions or implications.*

Don't be Wordy. Government writing is often noted for its wordiness. The following example is taken from a World War II price control regulation:

> "Ultimate consumer means a person or group of persons, generally constituting a domestic household, who purchase eggs generally at the individual stores or retailers or purchase and receive deliveries of eggs at the place of abode of the individual or domestic household from producers or retail route sellers and who use such eggs for their consumption or food."

This statement may be changed, without changing its meaning, by reducing it to:

"Ultimate consumers are people who buy eggs to eat them."

A good writer sees sharply what he wants to say, says directly what he sees, edits what he says, and takes pain to ease his reader's task. The example just given illustrates how much more readable a short sentence is than a long one.

7. WRITING A GOOD PARAGRAPH

The next major step after writing sentences is the construction of paragraphs. A good paragraph will clearly state the central ideas, it will fill in supporting details, and show how the central idea relates to the ideas which precede and follow it. The proper length of a paragraph is much like Abraham Lincoln's idea of a speech; it should be long enough to reach the end. There are exceptions, however. A report writer should be critical of his writing when a page of typewritten copy contains more than three or less than one complete paragraph.

Developing a Paragraph. There are many different ways of developing a paragraph in a report. The entry-level public-service worker may well utilize a variation of the following patterns of development:

- *Definition and description.* The writer gives his reader all the concepts he needs to follow the presentation.
- *Historical summary.* The writer briefs the reader on where the problem came from and why it is a problem. It provides a perspective for the problem.
- *Case history.* The report writer details an actual record for the reader.
- *Description of a process.* The report describes in detail how factors work together to produce a certain result.
- *Occasional summary.* The writer restates the essential data and ties facts together to clarify them for the reader.
- *Cause and effect.* The writer explains the forces that produce certain consequences.
- *Examination of alternatives.* The writer may present the material so that the reader can make an examination of alternatives and their possible consequences.
- *Directive.* The reader will be told what to do. The writer may describe steps to be taken.

Paragraph Construction. Regardless of the paragraph's development, a well-constructed paragraph will be correct, clear, and effective. Eight desirable paragraph characteristics are listed below:

- A good paragraph contains an implied or expressed topic sentence.
- A good paragraph is never sketchy or incomplete. It contains a complete body of thought.
- A good paragraph is mechanically correct. It is properly indented or otherwise set off. It correctly represents every change of speaker in dialogue.
- A good paragraph is unified. Extraneous details are eliminated.
- A good paragraph contains material arranged in proper order. Good arrangement of ideas demonstrates logical thinking on the part of the writer.
- A good paragraph should make orderly clear progress and there should be clear passage from one paragraph to another.
- A good paragraph will be of suitable length.
- A good paragraph is well proportioned to the importance of the content. The longest paragraph of the report should not deal with the least significant idea.

8. SPECIAL WORD FORMS

The report writer will need to know how to handle such other items as abbreviations and numbers, in addition to writing clear sentences in well constructed paragraphs.

Using Numbers Correctly. Nearly all reports contain numbers. Should the number be written out or should it be shown in figures? Many reports consist primarily of numbers. Such reports call for the use of figures. However, other reports which are not primarily numerical reports may leave the writer in doubt as to how the numbers should be shown. The following suggestions may be helpful:

- Numbers requiring the use of more than one word or a hyphenated word are usually written as figures.
- A number which begins a sentence is written as a word; if this is awkward or inconsistent with the rest of the text, replace the sentence.
- Numbers under ten are usually written as words unless they appear in writing which is full of numbers.

- Numbers that express dimensions are usually written as figures.
- Numbers which contain decimal fractions are always written as figures.
- Any number naming a common fraction is usually written as a hyphenated word.
- Numbers in any sentence in which other numbers are to occur in figure form are usually written in figures.
- Numbers that tell either time, date, or percentage are usually written in figures.
- Numbers above ten naming streets are usually written as figures with the indicators of pronounciation, -st, -nd, -th, attached.
- Round numbers, such as "thousands," are usually written as words except where numbers are occurring very frequently.
- Any number appearing occasionally and simple enough to express in one word is usually written as a word.
- Numbers appearing as two separate categories, one after the other, are usually written as words for the first category and as figures for the second. (five 2 x 4 boards)

The report writer should remember that the above guidelines are suggestions and that some of them may actually contradict each other. Common sense should determine whether or not a number be written as a figure or a word.

Using Abbreviations Correctly. Abbreviations are found more often in reports than in any other form of writing. Although the use of abbreviations is a healthy part of the style of field reports, it is definitely limited and kept in check by both tradition and common sense. *Only those abbreviations which will be easily understood by the reader are to be used.* Many agencies or departments will have a number of abbreviations that they use regularly, particularly in informal reports. Handbooks or manuals issued by the department will list these and the new worker should memorize them. The use of abbreviations has met with limited success because of the time saved in writing certain reports; the use should be very restricted in formal reports.

There are several rules regarding the general idea of permissible abbreviations in report writing:

- Whenever abbreviations have been used so long that they have assumed vocabulary status, they should be used. Some such abbreviations are Mr., Mrs., B.C., A.D., FBI, CIA, a.m., p.m. (Each job family will have a number of abbreviations of its own.)

- Whenever the names of units are preceded by numerals, it is usually best to abbreviate the names of the units. For instance:

    a. The auto was going 75-mph in the 25-mph zone.

    b. It was 65° F today.

    c. The unit has a 3-hp engine.

- If an abbreviation makes an English word (as for example, in. for inches), use a period. Otherwise, most good writers do <u>not</u> use periods in <u>most</u> cases.

If a term must be repeated many times in a report, that term calls for explanation and thereafter possible abbreviation, regardless of what it is.

<u>Using Contractions Correctly</u>. A contraction is a form of abbreviation. It is a word written with an apostrophe to indicate the omission of a letter. Contractions should be used very seldom in formal reports, but are common in semiformal and informal ones. Field reports are full of abbreviations and contractions as their use can greatly speed up the process of gathering data in the field.

<u>Using Capital Letters Correctly</u>. Report writing calls for no departure from the conventional rules for the use of capital letters. Proper names, names of cities and states, official titles, and organizations are always captalized. There are two practices common to reports:

- Capitalize all important words in titles, division headings, side headings, and captions. By "important" is meant all words except articles, prepositions, and conjunctions.
- Capitalize Figure, Table, Volume, Number as part of titles. Thus, reference would be made to Figure 4, Table 2, etc.

When in doubt, do not capitalize.

<u>Using Punctuation Correctly</u>. Clear communication is dependent upon yet another aspect of written language, punctuation. Every sentence begins with a capital letter, and ends with either a period, question mark, or exclamation point. Punctuation which does not contribute to the clarity of thought should be avoided. Most of the difficulties with punctuation arise out of the use of the comma, semicolon, and colon. For information on other punctuation, see any good handbook of grammar.

The Comma. The principal uses of the comma are:

- Between independent clauses connected by a coordinating conjunction (and, but, for, or, nor, yet). But if commas are used in any of the independent clauses constituting a sentence, a semicolon must be used between the clauses.
- After introductory clauses or phrases preceding the main clause of the sentence.
- Between items in a series.
- Around parenthetical phrases, appositives, and nonrestrictive modifiers.

The Semicolon. The semicolon is almost as strong a mark of separation as the period. It is chiefly used between the independent clauses not connected with one of the coordinating conjunctions, and between clauses connected with a coordinating conjunction which are long, or unrelated, or contain commas.

The Colon. The colon signals that something is to follow. It is a mark introducing lists, series, and quotations. It is used as a salutation in a business letter, in separating the hours and minutes in a statement of time, or in separating volume and pages in a bibliographical entry.

9. USING THE DICTIONARY IN REPORT WRITING

Dictionary Usage. The report writer should be aware of the fact that most dictionaries have a section listing the principal rules of capitalization, punctuation, and spelling. A dictionary should be considered one of the necessary tools for good report writing, together with the pencil and paper. The report writer should be using the dictionary for the following purposes:

- To determine the exact meaning of a word.
- To determine the correct spelling.
- To determine whether or not a word should be capitalized.
- To determine how a word should be divided at the end of a line.
- To determine correct pronunciation.
- To determine whether or not a hyphen should be used in a compound word.

Dictionary Content. In addition, the dictionary also has a list of common foreign words and phrases. It is obvious that much of the material that a report writer must master is to be found in a dictionary. The excellent report writer has the ability to select the exact word. No one can buy, sell, write letters, use the telephone, give orders, make a speech, or prepare a report, except by using words. Everything else being equal, the individual who knows the most about words will be the most successful in his occupation. *To develop the mastery of language necessary to use the exact word, the writer must know how to use the dictionary skillfully and he must use it frequently.*

The dictionary is a tool that will always be needed by the educated person. As a matter of fact, the better educated a person is, the more likely he is to refer frequently to a dictionary.

Word Division. One use of the dictionary mentioned above is for determining the correct place to divide a word at the end of a line. Since more errors are made in dividing words at the end of a line than in spelling them, capitalizing them, or in using them, it is important that the report writer learn to divide words correctly. Below are eight rules for dividing words at the end of a line:

- Never divide a word of one syllable.
- Do not divide a word of four letters.
- A one-letter syllable at the beginning of a word, or a one or two-letter syllable at the end of a word must not be separated from the rest of the word. (Examples: "about" not "a-bout"; "ready" not "read-y.")
- When a word containing three or more syllables is to be divided at a one-letter syllable, the one-letter syllable should be written on the first line rather than on the second. (Example: "maga-zine" not "mag-azine.")
- When a word is to be divided at a point where two vowels that are pronounced separately come together, these vowels should be divided into separate syllables. (Example: "continu-ation" not "continua-tion.") Note that this rule is an exception to the one stated above.
- A syllable that does not contain a vowel must not be separated from the remainder of the word. (Example: "wouldn't" not "would-n't.")
- Avoid dividing hyphenated words, such as "self-conscious," except at the hyphen.
- When a final consonant is doubled before a suffix, the additional consonant should be placed with the suffix. (Example: "run-ning" not "runn-ing.")

When in doubt about the proper syllable makeup of a word, consult the dictionary. Do not guess at the division of a word.

10. SUMMATION

English grammar essentials, such as correct punctuation, capitalization, syllabication, and correct use of numbers and abbreviations, are all part of the skills that the successful report writer has at his disposal. The beginning public-service worker is well advised to obtain a list of the words that appear frequently in his job family, or are a part of the technical or professional vocabulary, and memorize them. Common report forms should be reviewed and used as a guide for making observations and examinations of data when preparing to write a report. The writer may find that maps are needed to show geographical location, charts to visualize statistical data, or tables to determine relationships. Clarity will be the guide dictating how any idea may best be communicated to the reader.

In conclusion, the report writer must know what details to look for, must select the proper format for the report, must select the best possible words that do not have any emotional connotations and build them into short, effective sentences. Paragraphs must be developed by the writer around the central thoughts, leaving nothing to be imagined by the reader. The importance of word relationship and idea sequence is crucial.

Remember, a report is written to express an idea, not to impress a superior.

STUDENT LEARNING ACTIVITIES

- Write a report explaining a career choice.
- View the six filmstrips, *Constructing Reports*, and evaluate the information they contain.
- Complete the matching vocabulary exercise prepared by the teacher for the occupation group selected by the student.
- Demonstrate a knowledge of the active and passive verbs in report writing by changing a report prepared by the teacher from the passive voice to the active.
- List three words having unfavorable connotations and explain how they could be particularly embarrassing to the writer.
- Identify the type of paragraph development used in each of the paragraphs which the teacher has prepared.
- Keep a notebook for discussion notes and class handouts.
- Take a diagnostic test, and complete remedial lessons, if needed, on punctuation and capitalization.
- Complete a dictionary assignment.

- Prepare reports describing the events in a simulated classroom interruption.
- Prepare a group report on the students' attendance in class, citing the percentage of absentees each day. Compare Monday to Tuesday, etc., and list the frequency of reasons for the absences.
- Evaluate the reports prepared by the class members in the above two activities.

TEACHER MANAGEMENT ACTIVITIES

- Prepare a bulletin board display illustrating the components of a good report.
- Review the materials in the local audiovisual library to locate teaching aids that are readily available.
- Make arrangements for showing the movie, *Writing a Good Paragraph*.
- Make arrangements to show the filmstrips in the series: *Constructing Reports*. (6 filmstrips)
- Prepare discussion notes on the types of reports, and the importance and purpose of each type.
- Collect examples of reports from the eight job families in the Public Service Occupation area.
- Prepare a vocabulary list of technical and professional words from each of the public-service entry-level job families.
- Prepare matching exercises for the vocabulary words and their definitions for each of the major job groups.
- Prepare overlays for use with an overhead projector illustrating good and bad reports.
- Prepare a list of sentences containing passive-voice verbs which the students are to change to active verbs.
- Prepare a discussion of word meanings and emphasize the importance of connotation.
- Obtain or write, and present, paragraphs illustrating the seven ways of developing a paragraph.
- Plan simulated situations from which the students will have to prepare reports.
- Prepare a handout for students containing suggestions for the use of numbers in reports.
- Prepare discussion notes for the use of abbreviations in reports.
- Prepare a diagnostic quiz on punctuation marks and capitalization.

- Discuss and test the use of the dictionary in the classroom.

- Prepare a student handout on the eight rules for dividing words at the end of a line.

EVALUATION QUESTIONS

Basic Report Writing

1. **Public service workers are likely to write:**
   A. Formal reports
   B. Informal reports
   C. Semi-formal reports
   D. All of the above

2. **Which statement is untrue?**
   A. A good report writer does not have to be exact about facts
   B. A good report writer writes objectively and accurately about his observations
   C. A good report writer uses visual aids where they will help put over the message
   D. A good report writer gathers his material in an orderly way

3. **Police reports are important because:**
   A. They help settle arguments
   B. They help refresh the officer's memory
   C. Financial settlements may be involved
   D. All of the above

4. **Which statement is not true?**
   A. A good paragraph contains materials arranged in random order
   B. The length of the paragraph is suited to its importance
   C. A good paragraph is mechanically correct — indented or set off
   D. A good paragraph should make orderly, clear passage from one paragraph to another

5. **A good paragraph should:**
   A. Include all minor details
   B. Contain a topic sentence
   C. Leave out the main ideas
   D. Leave out the important details

6. **Inaccurate and incomplete reports can cause:**
   A. Misunderstanding
   B. Anger
   C. Costly errors
   D. All of the above

7. **Writers do not refer to dictionaries when they need to know:**
   A. Correct spelling
   B. How a word should be divided
   C. Detailed information about a subject
   D. Whether a hyphen should be used

8. Good report writers use:

A. Slang as much as possible ______
B. Words with unpleasant meanings ______
C. Phrases that rid the writer of responsibility ______
D. None of the above ______

9. When writing a report, it is best to use:

A. Phrases that are used over and over ______
B. Short, simple sentences ______
C. As many words as possible ______
D. An inactive tone ______

10. Which statement is incorrect?

A. A number which begins a sentence is written as a figure ______
B. Numbers under ten are usually written as words ______
C. Numbers requiring the use of more than one word are usually written as figures ______
D. Numbers which contain decimal fractions are always written as figures ______

11. Contractions are seldom used in:

A. Field reports ______
B. Semi-formal reports ______
C. Formal reports ______
D. Informal reports ______

12. Titles and the names of organizations are:

A. Never capitalized ______
B. Not capitalized in informal reports ______
C. Sometimes capitalized in formal reports ______
D. Always capitalized ______

13. After introductory clauses or phrases, one should use:

A. A period ______
B. A comma ______
C. A question mark ______
D. An exclamation mark ______

14. Between long, unrelated clauses that are connected with a conjunction, one should use:

A. A period ______
B. A question mark ______
C. A semi-colon ______
D. A colon ______

Answer Key

1. D	4. A	7. C	10. A	13. B
2. A	5. B	8. D	11. C	14. C
3. D	6. D	9. B	12. D	

THE NARRATIVE REPORT

CONTENTS

The Narrative Report

The narrative report is to an establishment what an introductory statement is to a definition. A good narrative report provides the context in which the particular study was made and serves to orient the reader to the circumstances under which jobs existed at that time. This background information is especially important to the reader before he examines the individual job analysis schedules. Further, the narrative report provides broad general occupational and industrial information which cannot be properly included in the job analysis schedule.

COMPILING THE DATA

The information in a narrative report grows out of discussions with workers, plant officials, industry experts, college or technical school personnel, and review of the technical literature pertaining to the industry. The analyst makes inquiries concerning the structure or organization of the establishment, the job interrelationships, the work flow processes, personnel policies and practices, hazards, and any other items which he feels will contribute to the job information. Much of this information can be obtained during the orientation tour, discussed in chapter III.

Examples of the kind of questions for which the analyst should seek answers are:

1. (If industrial) What is the purpose of the establishment? What are the processes by which raw materials are converted to finished products? (If non-industrial) What is the nature of the service rendered? What knowledge or technologies are required for adequate performance? How is the service rendered? What are the general duties or procedures?
2. What are the general environmental conditions? What are the hazards encountered by workers? What working conditions are peculiar to the establishment?
3. Are the services or work flow of the establishment divided into departments or units, and if so, how? How are these units interrelated, or how are they arranged in a work flow process?
4. What are the personnel practices? Are women used in other than traditional clerical jobs? Are minorities, handicapped, or older workers employed; if so, in what types of jobs? What is the trend in regard to such workers? Does a form of career lattice exist in this establishment? What are the entry jobs? Are training courses provided by the establishment?
5. Does this establishment have any unique characteristics in comparison to all other establishments in the industry? What is the history of the development of the establishment? Has it automated or initiated any progressive or unusual processes, equipment, or services? What effects have these new ideas or machinery had on the activities and employment status of the establishment?

II. ORGANIZING MATERIALS

Generally, the data obtained through the discussions with establishment personnel or other technical experts will fall fairly obviously under several major headings. Liberal use of such headings and their subheadings, even in a very short report, provides the reader with a ready reference to the particular sections of the report and an easier to read, more interesting text. In addition, the use of headings helps the analyst to organize his materials in terms of the headings and helps him set limits on the amount of information included in the narrative.

A standard formula for organizing the report is neither possible nor desirable, as each report and each analyst's presentation will include many different types of data. However, a few general headings have been found to be significant in most reports. The outline below presents these headings, with typical items explained in the contents. A report may not include all these headings, or it may include additional or other headings. These general headings are:

Introduction or purpose of establishment
- Purpose
- Scope and/or limit of study
- History and/or development of the establishment

Environmental Conditions
- Description of layout
- Description of equipment
- General environmental conditions and general working conditions

Organization and Operations or Activities
- Departmentalization of establishment
- Work flow
- Processes (if industrial)
- Services (if non-industrial)

Personnel Policies and Practices
- Hiring requirements
- Recruitment and/or sources of workers
- Methods of training
- Hiring of minorities, handicapped and disadvantaged
- Entry jobs
- Career lattices and/or promotional opportunities
- Job restructuring
- Effects of automation on personnel

Other Section (s)
- Comments
- Effects of automation on establishment or industry

Appendix and Glossary

Introduction or Purpose of the Establishment

The section under this heading should begin with a statement of the product or service of the establishment. If the study is in an industrial plant, this section would include a general discussion of the raw materials involved, the processes used, and the range and variety of products produced. If a non-industrial establishment is studied, the section would include a description of the nature of the service, and to whom and how the service is rendered. The name of the establishment does not appear on the report, which is identified by the establishment number only. This number appears at the top of the first page immediately beneath the words, Narrative Report.

Frequently, this section also includes a history of the establishment; what it began as, how it developed, and possibly some future plans of the establishment or future trends in the industry if these are significant to the employment status.

Any restrictions which the employer may impose on the study (such as processes which he wishes to keep secret) and which therefore affect the preparation of job analysis schedules, could be noted here.

Environmental Conditions

This section would include a description of the layout of the establishment, complete with size, to give the reader a picture of the physical arrangement of buildings, facilities, equipment, storage, or related areas as they affect work flow. The narration in this section would also include information on equipment, machines, and/or tools used, and working conditions. The description given here should be concerned with the overall establishment picture and should furnish information not contained in the individual job analysis schedules.

Organization and Activities

An explanation of the organization of the establishment gives the necessary orientation for reading the individual schedules. This section might begin with a discussion of the units, processes or major activities, and with their interrelation or position in the work flow. This would then be followed by a breakdown of the units, processes, or activities into separate subheadings with more detailed descriptions of each. When this section is read in conjunction with the environmental conditions section, it should enable the reader to visualize the total work situation into which the worker fits.

Personnel Policies and Practices

Here the analyst would give a picture of the establishment's hiring requirements and its methods of placement. Included under this heading are educational, physical, and other requirements; the employer's methods of recruitment; and his policy and practice in hiring members of special groups. In addition, subheadings deal with methods of entry and training given by or sanctioned by the establishment, and career lattices or other promotional opportunities which exist in the establishment.

If the purpose of the study is to find job-worker situations which could be restructured, or if the establishment has practiced some form of job restructuring, a discussion of this should appear here. Also, if the establishment has become automated and the automation has had significant impact on its employment, a subheading devoted to this should appear here. In either case, the discussion describes the effects of restructuring or automation on placement policies and practices, on employment statistics, and the resultant changes in educational and training requirements.

The following questions can serve as guides to the analyst for obtaining pertinent data:

Related to employment of women: What openings for women are there in this establishment? Are there environmental factors which negate successful employment? What is the establishment's attitude toward their employment? Have jobs been restructured to facilitate their employment?

Related to employment of the handicapped: Where and in what types of work are they employed? Is the establishment planning to accelerate the hiring of this type of worker? Has the possibility of restructuring jobs for them been considered?

Related to employment of members of minority groups: Are there limits to employment of these workers? To their promotional opportunities? Are there jobs or units in the establishment where minority workers are not employed? What is the employment trend in the hiring of these workers?

Related to employment of the disadvantaged worker: Does the establishment employ any less-than-fully qualified persons, or persons who qualify as disadvantaged? Has the establishment made any provision for special training for these workers? If not, could any jobs in the establishment be performed by a disadvantaged worker?

Related to training: Is there a formal in-plant training program? Are there outside facilities to which the establishment has recourse? Are there formal apprenticeships in effect?

Related to career lattices: Is there a formally established career lattice in the establishment? Are there well defined lines of promotion and transfer? Are in-step and on-the-job training programs integrated into this lattice? What are the entry occupations? What proportion of jobs in the establishment are entry occupations? Does the establishment participate in Federal training programs?

Related to job restructuring: Has this establishment tried any formal job restructuring? If not, could the jobs in this establishment be restructured? Are there factors which limit this possibility? What have been or what could be the results of restructuring in employment status? Has or could job restructuring create new or more entry occupations, and establish more meaningful career lattices?

Other Section (s)

Sometimes a report will include a section or sections devoted to special or unique topics.

For example, the analyst may include a section concerned with the product market, or special factors affecting work flow in the establishment. The discussion of the history of the establishment, or of future trends in the establishment or the industry may appear here rather than in the introduction or purpose of the establishment section.

A special case in this section is that concerned with the effects of automation or mechanization in an industry or establishment. In these instances a discussion of the effects of automation on personnel should appear in the personnel policies and practices section, and all other aspects of automation or mechanization, i.e., changes in equipment, changes in processes or activities, effects on working conditions and physical requirements, would be included in a special section at the end of the narrative report.

Appendix and Glossary

In the course of his study, the analyst may obtain from management or other sources supplemental materials such as brochures or forms, which might add additional interest and information to the report. These should be included as an appendix. The analyst may also find certain technical terms, processes, or equipment which he feels need to be clarified for the reader. These should appear in a glossary at the end of the report.

III. WRITING THE REPORT

Writing a narrative report is the process of converting the information secured into useable reference material. The report will present only pertinent and essential information in the fewest words consistent with clarity, proper arrangement for exposition, and accuracy of word usage.

Note: The SIC Code(s) which is identical to that entered on the staffing schedule should be entered on page one of the Narrative Report.

GUIDELINES FOR REPORT WRITING

1. The analyst should distinguish clearly between statements based on fact and those based on his own or other's opinions. At times, statements of opinions enhance the value of the narrative by rendering an overall picture of the study. However, the source of statements of opinion should be identified as, "In the analyst's opinion" or "The personnel manager states . . . ." Crediting a statement thus, while indicating that the statement is not substantiated completely, gives authority for the opinion and lends weight to it.
2. A paragraph must be built around one central thought, and sentences not contributing to that specific thought do not belong in the paragraph. However, breaks between paragraphs serve as resting points for the reader, and paragraphs of more than 200 or so words should not be used.
3. Suitable transition statements are necessary for the reader to follow the changing thought from paragraph to paragraph. Even when main headings and subheadings are used, the transition should be such that the reader understands that he has completed one thought and is progressing to the next.
4. Because of its position of emphasis, the opening sentence often is used to state the central thought which the remainder of the paragraph expands and supports. At other times it points the direction in which the new paragraph will move away from the preceding paragraph. The emphatic position of this first sentence in a paragraph, as well as the last sentence, should not be wasted by the writer.
5. The main headings may be centered and underlined. Secondary headings then can be placed at the left margin and underlined. Third-order headings (usually to be avoided) might be placed in the text and underlined. "Label" headings should be avoided. For example, while "Plant Environmental Factors" is adequate for a very broad heading, a subheading under it should say: "Physical Layout of Departments," rather than "Layouts." The format should be consistent.

An example of a narrative report follows.

EXAMPLE

NARRATIVE REPORT

SIC-2281

Establishment No. 360-150-392

Purpose of Establishment

This establishment is a processing plant within a synthetic yarn-producing division of a yarn and thread manufacturing corporation. It is engaged in spinning yarn from man-made fibers for use in manufacturing such articles as hosiery, pile fabrics, and men's and women's outerwear. Basic yarn counts produced range from 6's to 30's both single and plied.

Prior to 1959 this plant was engaged in the manufacture of carded cotton knitting yarns of coarse to medium count. The transition to producing synthetic yarns was completed in early 1959 with the installation of machinery developed for manufacturing synthetic yarn. This continuous processing system is a variation of the cotton processing system eliminating the processes which involved opening, cleaning, and transforming cotton fibers into laps preparatory to the carding process. This development is due to the fact that synthetic staples do not require extensive opening and cleaning as do natural fibers. The establishment has 9,792 spinning spindles, producing in excess of 175,000 pounds of synthetic yarn weekly.

Personnel Policies and Practices

Training for production jobs in this plant is usually on-the-job. Training periods extend from 2 weeks up to 2 years, the latter applicable to those persons engaged in setting up and repairing various machinery. No specialized training is required for entry jobs, only a general education being sufficient for communicating with coworkers and for learning the required tasks of the job. There are no definite lines of promotion; however, workers are upgraded into jobs that require more experience and skill as vacancies arise, based on their industriousness and willingness to assume responsibility. This is an equal opportunity establishment. There are no restrictions on the employment, training, and promotion of minority groups, women, or the handicapped. This establishment works three shifts.

Organization and Operations

The PLANT SUPERINTENDENT coordinates production activities for the plant. Subordinate supervisory personnel include a CARD ROOM SUPERVISOR, SPINNING ROOM SUPERVISOR, SHIPPING AND OUTSIDE FOREMAN, and MACHINIST. A SHIFT SUPERVISOR for both the second and third shifts works under the combined supervision of the CARD ROOM SUPERVISOR and SPINNING ROOM SUPERVISOR. This study was limited to the observation of production jobs.

Receiving

Synthetic fibers are shipped to the plant by manufacturers of man-made staples in boxes weighing up to 650 pounds. Fibers are unloaded from trucks and stored in the warehouse according to type. Boxes of synthetic staples are drawn from stock and positioned near blending machines with steel bands and tops removed to facilitate feeding fibers into machine hoppers.

Carding

This process as used in this establishment involves blending synthetic fibers and/or reusable waste, and feeding fibers through a distribution system into carding machines that produce sliver. Specified amounts of fibers fed into blending machines are deposited onto a conveyor from automatic weighing units attached to blending machines. Fibers are sprayed with fugitive dye tints and antistatic chemicals for identity as to type and to reduce friction in fibers during processing, and are conveyed through a piping system to automatic feeding units containing aprons with pins that feed fibers on a controlled basis to carding machines. Carding machines are equipped with several cylinders covered with metallic spikes that work in conjunction with carding drums to remove impurities from fibers, arrange fibers parallel, and produce sliver which is coiled in cans for use in the drawing process.

Drawing

This involves combining and passing several strands of sliver through two or more pairs of rollers, each of which rotate at a higher speed than the preceding pair, to attenuate the sliver.

Two phases of this process are used, namely breaker and finish drawing. In the initial step, eight slivers are fed into drawing machines that combine and straighten the fibers to produce a strand of uniform weight and size. The second phase combines eight breaker strands of sliver into one, thus improving the quality of the sliver processed. Sliver formed during the drawing process is coiled into cans for feeding into roving frames.

Roving

The purpose of this process is to combine and reduce sliver received from drawing frames into a continuous, slightly twisted strand called roving, and to wind roving onto bobbins for use in the spinning process. The drafting rollers of the roving frames draw out the sliver, and flyers slightly twist the roving as it is wound into bobbins.

Spinning

In this process ring spinning frames are used to reduce roving to yarn and to wind yarn onto bobbins. Roving from bobbins placed in the creels of spinning frames is drawn to its final size by sets of drafting rollers, twisted by travelers on the rings of spinning frames, and wound onto spinning bobbins.

Winding

This involves transferring yarn from spinning bobbins onto cones and spring coils through use of winding machines. A technique for joining broken ends together in specified yarn types is employed in addition to the use of hand knotters. This process, called "splicing" by management, involves gluing broken yarn ends together with a latex base compound, producing a knotless yarn.

Inspecting and Packing

Yarn packages are examined for finishing defects such as knots, soils, loose or tight winding, and absence of labels and specified color tip of cones. Ultraviolet lamps are used for detecting packages failing to meet blending specifications and for separating faultily mixed lots. Following inspection, yarn packages are wrapped in paper to prevent damage to yarn during shipment, and packed in shipping cartons. Cartons are stenciled with identifying information, weighed, strapped with steel bands, and moved to the shipping area by a conveyor.

Shipping

Customer order shipments are loaded onto trucks and transported to a central warehouse for consolidation of orders and delivery to customers of subsidiary plants, following priority of orders.

Environmental Conditions

The physical plant was constructed during the 1920's, but has been remodeled in the past 7 years. It is adequately lighted and ventilated and is clean. A cafeteria, containing a coin-operated food-and-beverage dispensing machine, is available for use by workers during breaks and lunch periods. Smoking areas are also provided and so designated to minimize fire hazards.

The noise level is considered critical as a result of the constant operation of machines throughout the plant. Automatic vacuum piping systems and overhead traveling cleaners reduce the amount of lint and other foreign matter in the carding and spinning rooms that could result in worker discomfort.

Workers handling cartons of yarn and fibers or other heavy objects work together as team members or use lifting devices and handtrucks in moving materials and supplies to prevent personal injury. Though the possibility for injuries exists for personnel working with or around machines, strict observance of safety rules and regulations rarely results in serious injury. Selected workers from each shift are trained in rendering first-aid treatment when minor injuries occur.

Special Comments

The processing of synthetic fibers into sliver using the system outlined eliminates the picking process which involves transforming fibers into laps for use in the carding process. The replacement of revolving flats with a series of rolls containing metallic spikes reduces the grinding and stripping operations usually accompanying the carding of natural fibers as synthetic fibers do not require extensive cleaning. As a result, such tasks as feeding blending machines and tending carding machines have been added to the carding process.

Glossary

6's to 30's: coarse to medium yarn.

Hoppers: Units containing aprons with spikes that remove compression from synthetic fibers. Several types of fibers and waste can be blended into this unit with unusable waste fibers removed by piping system.

Sliver: Loose, untwisted strand of synthetic fibers produced on carding machines and drawing frames.

Fugitive Dye: A dye which is not fast.

Attenuate: To make slender or thin.

Cans: Large, cylindrical containers used to receive and hold sliver delivered from drawing frames for feeding into roving frames.

Drafting Rollers: Two or more pairs of rollers, each pair of which rotates at a higher speed than the preceding pair, serving to attenuate the roving passing between them.

Carding Machine: Machine used to remove impurities from synthetic fibers, arrange fibers parallel, and produce sliver for drawing process. The machine consists of several cylinders covered with metallic teeth that card the fibers.

Drawing Frame: Machine used to combine several strands of sliver and draw out strand to produce one of uniform weight and size.

Spinning Frame: Machine used to draw out and transform slightly twisted roving into yarn and wind yarn onto bobbins.

Roving Frame: Machine used to draw out strands of sliver and loosely twist them together to form roving.

Winding Machine: Machine used to transfer yarn from bobbins onto cones and spring coils.

Traveler: A small free-running metal ring sliding on a bar through which thread passes into other textile machine to impart a twist to the thread.

ANSWER SHEET

EST NO. ______________ PART ______ TITLE OF POSITION ______________________________

(AS GIVEN IN EXAMINATION ANNOUNCEMENT - INCLUDE OPTION, IF ANY)

LACE OF EXAMINATION ______________________________ ______________________________ DATE ______________

(CITY OR TOWN) (STATE)

RATING

USE THE SPECIAL PENCIL. MAKE GLOSSY BLACK MARKS.

1	A B C D E	26	A B C D E	51	A B C D E	76	A B C D E	101	A B C D E
2	A B C D E	27	A B C D E	52	A B C D E	77	A B C D E	102	A B C D E
3	A B C D E	28	A B C D E	53	A B C D E	78	A B C D E	103	A B C D E
4	A B C D E	29	A B C D E	54	A B C D E	79	A B C D E	104	A B C D E
5	A B C D E	30	A B C D E	55	A B C D E	80	A B C D E	105	A B C D E
6	A B C D E	31	A B C D E	56	A B C D E	81	A B C D E	106	A B C D E
7	A B C D E	32	A B C D E	57	A B C D E	82	A B C D E	107	A B C D E
8	A B C D E	33	A B C D E	58	A B C D E	83	A B C D E	108	A B C D E
9	A B C D E	34	A B C D E	59	A B C D E	84	A B C D E	109	A B C D E
10	A B C D E	35	A B C D E	60	A B C D E	85	A B C D E	110	A B C D E

Make only ONE mark for each answer. Additional and stray marks may be counted as mistakes. In making corrections, erase errors COMPLETELY.

11	A B C D E	36	A B C D E	61	A B C D E	86	A B C D E	111	A B C D E
12	A B C D E	37	A B C D E	62	A B C D E	87	A B C D E	112	A B C D E
13	A B C D E	38	A B C D E	63	A B C D E	88	A B C D E	113	A B C D E
14	A B C D E	39	A B C D E	64	A B C D E	89	A B C D E	114	A B C D E
15	A B C D E	40	A B C D E	65	A B C D E	90	A B C D E	115	A B C D E
16	A B C D E	41	A B C D E	66	A B C D E	91	A B C D E	116	A B C D E
17	A B C D E	42	A B C D E	67	A B C D E	92	A B C D E	117	A B C D E
18	A B C D E	43	A B C D E	68	A B C D E	93	A B C D E	118	A B C D E
19	A B C D E	44	A B C D E	69	A B C D E	94	A B C D E	119	A B C D E
20	A B C D E	45	A B C D E	70	A B C D E	95	A B C D E	120	A B C D E
21	A B C D E	46	A B C D E	71	A B C D E	96	A B C D E	121	A B C D E
22	A B C D E	47	A B C D E	72	A B C D E	97	A B C D E	122	A B C D E
23	A B C D E	48	A B C D E	73	A B C D E	98	A B C D E	123	A B C D E
24	A B C D E	49	A B C D E	74	A B C D E	99	A B C D E	124	A B C D E
25	A B C D E	50	A B C D E	75	A B C D E	100	A B C D E	125	A B C D E

ANSWER SHEET

TEST NO. ________ PART ________ TITLE OF POSITION ________
(AS GIVEN IN EXAMINATION ANNOUNCEMENT - INCLUDE OPTION, IF ANY)

PLACE OF EXAMINATION ________ (CITY OR TOWN) ________ (STATE) DATE ________

RATING

USE THE SPECIAL PENCIL. MAKE GLOSSY BLACK MARKS.

1 A B C D E	26 A B C D E	51 A B C D E	76 A B C D E	101 A B C D E
2 A B C D E	27 A B C D E	52 A B C D E	77 A B C D E	102 A B C D E
3 A B C D E	28 A B C D E	53 A B C D E	78 A B C D E	103 A B C D E
4 A B C D E	29 A B C D E	54 A B C D E	79 A B C D E	104 A B C D E
5 A B C D E	30 A B C D E	55 A B C D E	80 A B C D E	105 A B C D E
6 A B C D E	31 A B C D E	56 A B C D E	81 A B C D E	106 A B C D E
7 A B C D E	32 A B C D E	57 A B C D E	82 A B C D E	107 A B C D E
8 A B C D E	33 A B C D E	58 A B C D E	83 A B C D E	108 A B C D E
9 A B C D E	34 A B C D E	59 A B C D E	84 A B C D E	109 A B C D E
10 A B C D E	35 A B C D E	60 A B C D E	85 A B C D E	110 A B C D E

Make only ONE mark for each answer. Additional and stray marks may be counted as mistakes. In making corrections, erase errors COMPLETELY.

11 A B C D E	36 A B C D E	61 A B C D E	86 A B C D E	111 A B C D E
12 A B C D E	37 A B C D E	62 A B C D E	87 A B C D E	112 A B C D E
13 A B C D E	38 A B C D E	63 A B C D E	88 A B C D E	113 A B C D E
14 A B C D E	39 A B C D E	64 A B C D E	89 A B C D E	114 A B C D E
15 A B C D E	40 A B C D E	65 A B C D E	90 A B C D E	115 A B C D E
16 A B C D E	41 A B C D E	66 A B C D E	91 A B C D E	116 A B C D E
17 A B C D E	42 A B C D E	67 A B C D E	92 A B C D E	117 A B C D E
18 A B C D E	43 A B C D E	68 A B C D E	93 A B C D E	118 A B C D E
19 A B C D E	44 A B C D E	69 A B C D E	94 A B C D E	119 A B C D E
20 A B C D E	45 A B C D E	70 A B C D E	95 A B C D E	120 A B C D E
21 A B C D E	46 A B C D E	71 A B C D E	96 A B C D E	121 A B C D E
22 A B C D E	47 A B C D E	72 A B C D E	97 A B C D E	122 A B C D E
23 A B C D E	48 A B C D E	73 A B C D E	98 A B C D E	123 A B C D E
24 A B C D E	49 A B C D E	74 A B C D E	99 A B C D E	124 A B C D E
25 A B C D E	50 A B C D E	75 A B C D E	100 A B C D E	125 A B C D E